普通高等教育"十二五"规划教材

计算机网络实践指导教程

王 涛 裘国永 主编

科学出版社

北京

内 容 简 介

本书包含网络协议分析和套接字编程实践两部分，其中网络协议分析围绕计算机网络中 HTTP、DNS、DHCP、TCP、UDP、IP、ICMP、NAT、Ethernet、ARP、802.11 等协议展开，设计编写了 11 套 Wireshark 协议分析实验，带领读者通过分析报文格式和运行中的协议交互过程，深入了解网络协议的设计和工作原理。套接字编程实践以 TCP 和 UDP 套接字编程为基础，设计编写了 9 个编程实验，由浅入深、由易到难地覆盖了 Web 服务器、邮件客户端、Web 代理、可靠传输、路由算法以及多媒体点播等内容，涉及网络体系结构的多个层次。代码实现利用 JAVA 语言和 C 语言两大主流工具，带领读者了解、熟悉、精通套接字编程方法。

本书可作为高等院校本科生和研究生计算机网络实践课的教材，也可以作为科研和工程技术人员的参考资料。

图书在版编目(CIP)数据

计算机网络实践指导教程/王涛，裘国永主编．—北京：科学出版社，2015.9

普通高等教育"十二五"规划教材
ISBN 978-7-03-045890-2

Ⅰ．①计… Ⅱ．①王… ②裘… Ⅲ．①计算机网络-高等学校-教材
Ⅳ．①TP393

中国版本图书馆 CIP 数据核字(2015)第 234411 号

责任编辑：李 萍 杨向萍 纪四稳/责任校对：钟 洋
责任印制：赵 博/封面设计：红叶图文

科学出版社 出版
北京东黄城根北街 16 号
邮政编码：100717
http://www.sciencep.com

源海印刷厂印刷
科学出版社发行 各地新华书店经销

*

2015 年 10 月第 一 版 开本：720×1000 1/16
2015 年 10 月第一次印刷 印张：15 1/2
字数：310 000
定价：58.00 元
（如有印装质量问题，我社负责调换）

前 言

计算机网络对于相关专业的从业人员是必备知识。尤其在当今"互联网＋"的大背景下,互联网思维已经深入人心,如何更好地利用计算机网络的相关知识去解决实际问题是学习这门学科最主要的任务。然而,计算机网络是一门内容复杂的学科,涉及纷繁复杂的概念、协议和技术,同时还具备很强的实践性,因此仅凭理论学习是不足以深入了解网络协议背后的设计和工作原理,更谈不上去设计和开发一个网络应用程序。

作者从 2006 年开始从事计算机网络和网络工程等课程的教学工作,经过多年的积累形成了本书协议分析和套接字编程的讲义和素材,并已在多年的内部教学使用过程中得到了检验。因此,给计算机网络的学习者提供实践指导,并对作者多年的教学实践工作进行总结正是编写本书的初衷。

本书分为两部分。第一部分是网络协议分析部分,以 TCP/IP 体系为线索,自顶向下地由应用层到运输层,再到网络层和数据链路层,针对最常见的一些协议如 HTTP、DNS、DHCP、TCP、UDP、IP、ICMP、NAT、Ethernet、ARP、802.11 等展开分析,并以 Wireshark 为主要工具提供实验设计和指导。带领读者通过分析协议报文格式和运行中的交互过程,深入了解网络协议的设计和工作原理。内容包括:第 1 章协议分析工具,第 2 章应用层典型协议分析,第 3 章运输层典型协议分析,第 4 章网络层典型协议分析,第 5 章数据链路层和局域网典型协议分析。

第二部分以 Kurose 等编写的国际著名教材 *Computer Networking: A Top-Down Approach* 第 6 版提供的编程作业为线索,利用 JAVA 语言和 C 语言两大主流工具设计 9 个编程实验,内容覆盖 Web 服务器、邮件客户端、Web 代理、可靠传输、路由算法以及多媒体点播等。带领读者了解、熟悉、精通套接字编程方法。内容包括:第 6 章 TCP 和 UDP 套接字编程,第 7 章多线程 Web 服务器,第 8 章邮件客户端,第 9 章邮件用户代理:控制台版本,第 10 章用 UDP 实现 ping 功能,第 11 章 Web 代理服务器,第 12 章实现一个可靠传输协议,第 13 章一个分布式异步距离向量算法,第 14 章 RTSP 和 RTP 实现流媒体点播系统。

书中的理论内容由王涛、裴国永、吴振强等负责编写。第 1、3 章实验由秦石醉负责编写;第 2、4、5 章实验由张阳阳负责编写;第 6～14 章实验由张阳阳、秦石醉、王涛负责编写。同时感谢张夏蕾、黄亚军等在套接字编程部分源代码实现等方面的工作。全书由王涛负责设计、规划、统稿。本书相关电子资源可参考 http://netresearch.snnu.edu.cn。

  本书的编写得到了陕西师范大学计算机科学学院的大力支持，得到了王小明、马苗等许多同事的指导和支持，也得到了陕西师范大学校级优秀教材出版项目以及陕西师范大学基地班、创新实验班专项建设项目的资助，作者在此一并表示衷心的感谢！

  教材编写是在不断的教学实践中摸索总结出来的，由于作者水平有限，书中难免有一些不足之处，恳请读者多提宝贵意见，给予批评指正，不胜感激！

作　者

2015 年 7 月于西安

目 录

前言

第一部分 网络协议分析

第1章 协议分析工具 ······ 3
  1.1 协议分析及工具 ······ 3
  1.2 下载 Wireshark ······ 5
  1.3 运行 Wireshark ······ 6
  1.4 Wireshark 过滤条件表达式 ······ 7
  1.5 使用 Wireshark 进行测试 ······ 8

第2章 应用层典型协议分析 ······ 9
  2.1 网络应用程序的工作模式 ······ 9
  2.2 超文本传输协议 ······ 10
  2.3 域名系统 ······ 20
  2.4 动态主机配置协议 ······ 33

第3章 运输层典型协议分析 ······ 42
  3.1 运输层概述 ······ 42
  3.2 TCP ······ 42
  3.3 UDP ······ 54

第4章 网络层典型协议分析 ······ 59
  4.1 网络层简介 ······ 59
  4.2 网际协议 IPv4 ······ 62
  4.3 互联网控制消息协议 ······ 74
  4.4 网络地址转换 ······ 81

第5章 数据链路层和局域网典型协议分析 ······ 87
  5.1 数据链路层的概述和服务 ······ 87
  5.2 以太网协议 ······ 88
  5.3 地址解析协议 ······ 93
  5.4 无线局域网协议 802.11 ······ 97

第二部分　套接字编程实践

第 6 章　TCP 和 UDP 套接字编程 …………………………………………… 109
- 6.1　什么是套接字 …………………………………………………………… 109
- 6.2　套接字的属性 …………………………………………………………… 109
- 6.3　服务器端与客户端 ……………………………………………………… 110
- 6.4　运输层套接字的使用 …………………………………………………… 111
- 6.5　Windows 平台 TCP 套接字的接口及使用 …………………………… 111
- 6.6　TCP 套接字编程 ………………………………………………………… 114
- 6.7　UDP 套接字编程 ………………………………………………………… 117

第 7 章　多线程 Web 服务器 ………………………………………………… 120
- 7.1　实验目标 ………………………………………………………………… 120
- 7.2　系统设计与组成 ………………………………………………………… 120
- 7.3　重要类及方法 …………………………………………………………… 120
- 7.4　开发环境 ………………………………………………………………… 121
- 7.5　运行结果 ………………………………………………………………… 121
- 7.6　源代码 …………………………………………………………………… 122

第 8 章　邮件客户端 …………………………………………………………… 128
- 8.1　实验目标 ………………………………………………………………… 128
- 8.2　系统设计与组成 ………………………………………………………… 129
- 8.3　重要类及方法 …………………………………………………………… 129
- 8.4　开发环境 ………………………………………………………………… 129
- 8.5　运行结果 ………………………………………………………………… 130
- 8.6　源代码 …………………………………………………………………… 131

第 9 章　邮件用户代理：控制台版本 ………………………………………… 142
- 9.1　实验目标 ………………………………………………………………… 142
- 9.2　系统设计与组成 ………………………………………………………… 142
- 9.3　重要的类及实现 ………………………………………………………… 143
- 9.4　开发环境 ………………………………………………………………… 143
- 9.5　运行结果 ………………………………………………………………… 143
- 9.6　源代码 …………………………………………………………………… 144

第 10 章　用 UDP 实现 ping 功能 …………………………………………… 147
- 10.1　实验目标 ………………………………………………………………… 147
- 10.2　系统设计与组成 ………………………………………………………… 147
- 10.3　重要的类及实现 ………………………………………………………… 147

10.4　开发环境 ··· 148
　　10.5　运行结果 ··· 148
　　10.6　源代码 ··· 150
第 11 章　Web 代理服务器 ··· 154
　　11.1　实验目标 ··· 154
　　11.2　系统设计与组成 ··· 154
　　11.3　重要类及方法 ··· 155
　　11.4　开发环境 ··· 155
　　11.5　运行结果 ··· 155
　　11.6　源代码 ··· 157
第 12 章　实现一个可靠传输协议 ····································· 167
　　12.1　实验目标 ··· 167
　　12.2　系统设计与组成 ··· 167
　　12.3　重要方法 ··· 169
　　12.4　开发环境 ··· 170
　　12.5　运行结果 ··· 170
　　12.6　源代码 ··· 172
第 13 章　一个分布式异步距离向量算法 ······························· 187
　　13.1　实验目标 ··· 187
　　13.2　系统设计与组成 ··· 187
　　13.3　系统设计 ··· 187
　　13.4　重要方法 ··· 188
　　13.5　开发环境 ··· 188
　　13.6　运行结果 ··· 189
　　13.7　源代码 ··· 190
第 14 章　RTSP 和 RTP 实现流媒体点播系统 ··························· 210
　　14.1　实验目标 ··· 210
　　14.2　系统设计与组成 ··· 210
　　14.3　重要类及方法 ··· 211
　　14.4　开发环境 ··· 212
　　14.5　运行结果 ··· 212
　　14.6　源代码 ··· 214
参考文献 ··· 238

第一部分

网络协议分析

第1章 协议分析工具

1.1 协议分析及工具

1.1.1 协议分析概述

在协议分析实验中，需要借助协议分析软件对数据进行抓包观察和分析。观察正在执行协议的两个实体间报文交互的基本工具称为分组嗅探器，一个分组嗅探器俘获（嗅探）计算机发送和接收的报文。一般情况下，分组嗅探器将存储和显示出被俘获报文的各协议首部字段的内容，它的作用相当于对报文和协议进行复制以便分析。

图1-1显示了分组嗅探器的结构。右侧是协议和应用程序，协议分为5层，自上而下分别是应用层、运输层、网络层、数据链路层和物理层，在数据链路层帧中包含了以上所有协议层的报文。左侧数据分组抓包（嗅探）器分为两部分：分组分析和分组抓包。分组分析器在协议下方显示了该协议所有的字段，分组抓包在其抓包库中存储了计算机发送和接收数据链路层帧副本。

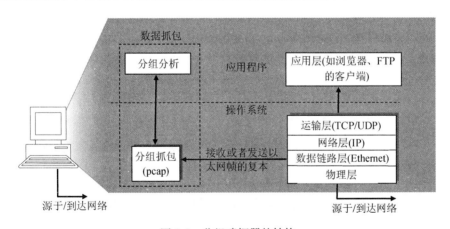

图1-1 分组嗅探器的结构

1.1.2 协议分析工具

为了深入地理解网络协议，观察运行中的协议行为以及分析协议的交互过程是最好的途径。通过对协议使用实体间交互报文序列的观察，研究协议操作细节并且使用协议执行确切的操作及分析其结果。为此，需要利用协议分析工具进行协议分析实验。

目前常见的网络数据抓包工具有 Wireshark、FireFox 里面的 HttpFox、Internet Explorer 里面的 HttpWatch、Fiddler2 和 SmartSniff 等。

HttpFox 的运行界面如图 1-2 所示。

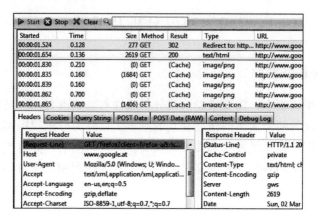

图 1-2　HttpFox 运行界面

HttpWatch 运行界面如图 1-3 所示。

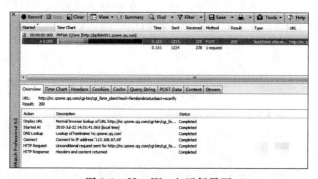

图 1-3　HttpWatch 运行界面

SmartSniff 的运行界面如图 1-4 所示。

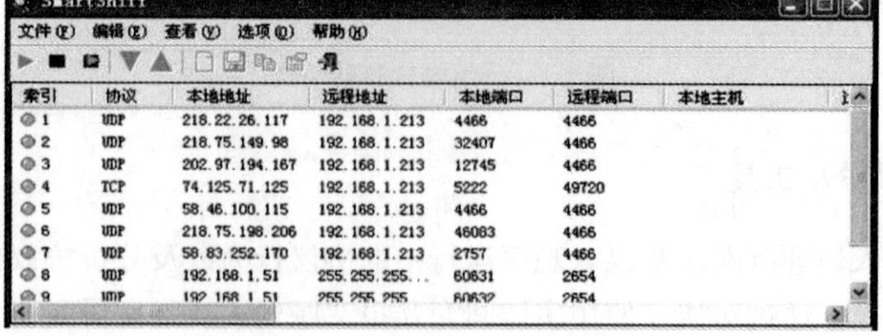

图 1-4　SmartSniff 运行界面

Wireshark 运行界面如图 1-5 所示。

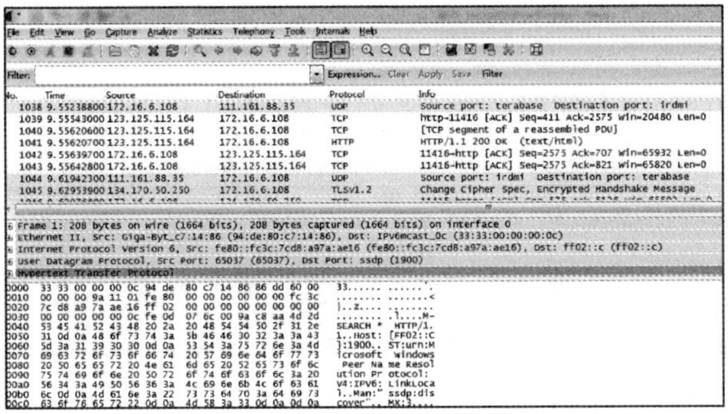

图 1-5　Wireshark 运行界面

其中，Wireshark 是免费软件，运行稳定，是目前最主流的开源网络分析软件，成为本书协议分析实验的首选。下面对 Wireshark 软件的发展历史进行简单介绍。

1.1.3　Wireshark 发展历史

1997 年年底，Combs 需要一个能够追踪网络流量的工具软件作为其工作上的辅助。因此，他开始编写 Ethereal 软件。Ethereal 在经过几次中断开发的事件过后，终于在 1998 年 7 月发布了第一个版本 V0.2.0。自此之后，Combs 收到了来自世界各地的修补程序、错误回报与鼓励信件。Ethereal 的发展就此开始。不久之后，Ramirez 看到了这套软件的开发潜力并开始参与开发。1998 年 10 月，来自 Network Appliance 公司的 Harris 在寻找一套比 TCPView（另外一个分组嗅探）更好的软件。于是，他也开始参与 Ethereal 的开发工作。1998 年底，一位教授 TCP/IP 课程的讲师 Sharpe 看到了这套软件的发展潜力，而后开始参与开发与加入新协议的功能。在当时，新的通信协议的制订并不复杂，因此他开始在 Ethereal 上新增分组嗅探功能，几乎包含了当时所有通信协议。

自此之后，数以千计的人开始参与 Ethereal 的开发，2006 年 6 月，因为商标的问题，Ethereal 更名为 Wireshark。

1.2　下载 Wireshark

为了进行 Wireshark 实验，需要下载 Wireshark 软件以及 libpcap 或者 WinPCap 分组抓包库。可访问 http：//www.Wireshark.org/download.html 选择支持的操作系统以便下载。

Wireshark 的 FAQ 有大量的有用的提示，尤其是在安装软件过程中遇到问题也可以在其中找到答案。

1.3 运行 Wireshark

当运行 Wireshark 程序时，进入开始界面如图 1-6 所示。

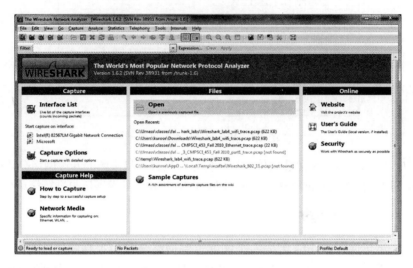

图 1-6 初始的 Wireshark 界面

观察界面左上角将会看到"接口列表"（Interface List），这是使用者计算机上网络接口的列表，一旦选择一个接口，软件将抓取该接口所有的数据分组。选中一个接口并开始数据抓包，屏幕将显示如图 1-7 所示信息，在 Capture 的下拉菜单可以选择停止数据抓包。

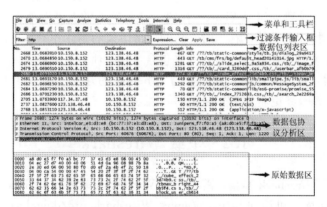

图 1-7 Wireshark 用户界面数据抓包分析

Wireshark 界面有 5 个主要部分。

（1）菜单和工具栏。在窗口顶部，有标准的下拉菜单。主要关注 File 和 Capture 菜单。File 菜单有允许存取数据分组以及打开之前抓取的分组数据,以及退出应用程序等功能。Capture 菜单有执行开始数据抓包等功能。

(2) 过滤条件输入框。在此可以使用多种表达式进行筛选和过滤。

(3) 数据包列表区。该区域是每一个抓取到数据分组的摘要信息,包括分组包的序号(由 Wireshark 分配)、分组抓取的时间、分组的源地址和目的地址、协议类型以及在分组中的特定协议信息。分组列表可以通过单击列名进行排序。协议类型显示的是发送或者接收该协议的最上层协议。

(4) 数据包协议分析区。提供了选定分组的详细信息,这些详细信息包括以太帧(假定数据分组是通过以太网发送或者接收的)和 IP 数据报。以太帧和 IP 层可以通过单击前面的"+"展开或者"-"收起。如果分组通过 TCP 或者 UDP 运输,TCP 或 UDP 详情也会显示。在该窗口默认显示选定分组的最上层协议。

(5) 原始数据区。显示的是捕获分组的原始二进制信息,包括左边的十六进制格式以及右边的 ASCII 码格式。

1.4　Wireshark 过滤条件表达式

1.4.1　针对 IP 地址的过滤

Wireshark 最常用的是针对 IP 地址的过滤,其中有以下几种情况。

(1) 对源地址过滤。例如,对源地址为 192.168.0.1 的包的过滤,即抓取源地址满足要求的包,表达式为 ip.src == 192.168.0.1。

(2) 对目的地址过滤。例如,对目的地址为 192.168.0.1 的包的过滤,即抓取目的地址满足要求的包,表达式为 ip.dst == 192.168.0.1。

(3) 对源或者目的地址过滤。例如,对源或者目的地址过滤为 192.168.0.1 的包的过滤,即抓取满足源或者目的地址的 IP 地址是 192.168.0.1 的包,表达式为 ip.addr == 192.168.0.1,或者 ip.src == 192.168.0.1 or ip.dst == 192.168.0.1。

(4) 要排除以上的数据包,只需要将其用括号囊括,然后使用"!"即可。表达式为!。

1.4.2　针对协议的过滤

(1) 仅仅需要捕获某种协议的数据包,表达式仅需要把协议的名字输入即可,如表达式为 http。

(2) 需要捕获多种协议的数据包,也只需对协议进行逻辑组合即可,如表达式为 http or telnet(多种协议加上逻辑符号的组合即可)。

(3) 排除某种协议的数据包。表达式用"!"表示,如表达式为 not arp、!tcp 等。

1.4.3 针对端口的过滤（视协议而定）

（1）捕获某一端口的数据包，表达式为 tcp.port == 端口号，如 tcp.port == 80。

（2）捕获多端口的数据包，可以使用 and 来连接，下面是捕获高端口的表达式，表达式为 udp.port>=端口号，如 udp.port>= 2048。

1.4.4 针对长度和内容的过滤

（1）针对长度的过滤（长度指定的是数据段的长度），表达式为协议报文长度+"关系算符"+长度值，如 udp.length< 30、http.content _ length<=20。

（2）针对数据包内容的过滤，如表达式为 http.request.uri matches "vipscu"（匹配 http 请求中含有 vipscu 字段的请求信息）。

1.5　使用 Wireshark 进行测试

假定计算机通过有线以太网连接到网络，按如下步骤抓包。

（1）打开浏览器，将会显示选择的主页。

（2）打开 Wireshark 软件，将看到如图 1-6 所示的初始窗口，此时软件还没有进行数据抓包。

（3）在 Capture 下拉菜单选择接口 interface，会显示窗口："Wireshark：Capture Interfaces"，如图 1-8 所示。

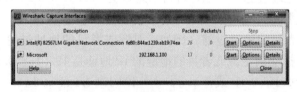

图 1-8　选择抓包接口对话框

（4）单击 Start，软件开始数据抓包，它将记录该接口所有接收和发送的数据分组。在软件运行时，在浏览器地址栏输入 www.snnu.edu.cn，按回车键运行，当跳到指定页面时选择停止数据抓包。Wireshark 软件记录了从开始以后在选定的接口中，到按 Stop，所有的接收和发送的数据分组。

（5）在协议过滤框输入 http、dns 或者其他信息，以便选择需要的信息进行选择和分析。

（6）退出 Wireshark。

在对数据抓包工具 Wireshark 及其简单的操作方法做了初步了解后，下面就可以开始本书的实验。

第 2 章 应用层典型协议分析

应用层协议定义了网络应用程序进程之间的通信规范。每个应用层协议都是为了解决某一类应用问题而出现的，例如，用于 WWW 应用的 HTTP 协议、用于文件传输的 FTP 协议、用于邮件传输的 SMTP 协议以及实现域名解析服务的 DNS 协议等。

本章首先将介绍网络应用程序的工作模式，接着重点对应用层中的 HTTP、DNS 以及 DHCP 等协议进行分析。

2.1 网络应用程序的工作模式

2.1.1 客户/服务器模式

在互联网中，最主要的进程间交互的方式是客户/服务器模式，即 Client/Server，简称 C/S 模式。在客户/服务器体系结构中，有一个（或多个）总是开机且在线的主机称为服务器，它服务于来自许多其他称为客户机的主机请求。客户机主机可能有时打开，也可能总是打开。一个典型的例子是 Web 应用程序，其中总是打开的 Web 服务器为来自运行在客户机上的浏览器的请求提供服务。在这里需要注意的是，C/S 模式中的客户机之间不直接通信。例如，在 Web 应用中，两个浏览器之间并不直接通信。C/S 体系结构的另一个特征是服务器有固定的、周知的地址，称为 IP 地址。因为服务器具有固定的、周知的地址，并且总是处于打开的状态，所以客户机总是能够向该服务器的地址发送分组来与其联系。图 2-1 显示了这种 C/S 体系结构。

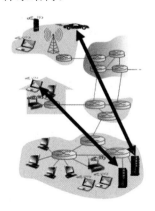

图 2-1 网络应用程序的 C/S 体系结构

客户发出请求完全是随机的，可能会有两个甚至多个请求同时到达同一个服务器，这时就会发生一台服务器主机无法满足其所有客户机请求的情况。为此，在C/S体系结构中，常用主机群集（有时称为服务器场，位于数据中心）创建强大的虚拟服务器。

2.1.2　P2P模式

任意间歇性连接的主机对称为对等方，可直接相互通信，对等方通信不必通过专门的服务器，所以该体系结构称为对等方到对等方，即Peer to Peer模式，简称P2P模式。它们通过直接交换信息达到共享计算机资源和服务的目的。目前，P2P技术已广泛应用于实时通信、协同工作、内容分发与分布式计算等领域。统计数据表明，目前的互联网流量中P2P流量已超过60%，已经成为当前互联网应用的新的重要形式，也是当前网络技术研究的热点问题之一。

P2P模式中，最突出的特征之一是它的自扩展性。例如，在一个P2P文件共享应用中，尽管每个对等方都由请求文件产生负载，但每个对等方向其他对等方分发文件的同时也为系统增加了服务能力。但另外，由于P2P应用程序具有高度分布和开放的性质，所以要格外关注系统的安全。P2P模式的应用环境如图2-2所示。

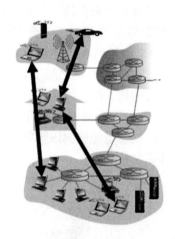

图2-2　网络应用程序的P2P工作模式

2.2　超文本传输协议

WWW（World Wide Web）是目前最常见的Internet服务，它用超文本作为文档的标准格式。超文本传输协议（HTTP）则负责把超文本从Web服务器传输到客户端。

2.2.1 HTTP 简介

HTTP（hypertext transfer protocol），即超文本传输协议，是一种 Internet 上最常见的协议，是 Web 浏览器和 Web 服务器之间的应用层通信协议，用于传输超文本标记语言（HTML）编写的文件，也就是通常说的网页。通过 HTTP，可以浏览网络上的各种信息，在浏览器上看到丰富多彩的文字与图片。HTTP 是基于 TCP/IP 之上的协议，它不仅保证正确传输超文本文档，还确定传输文档中的哪一部分，以及哪一部分内容首先显示（如文本先于图形）等。HTTP 基本的交互模式如图 2-3 所示。

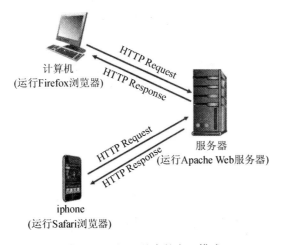

图 2-3 HTTP 基本的交互模式

HTTP 使用 TCP（而不是 UDP）作为它的支撑运输层协议。HTTP 客户机发起一个和服务器的 TCP 的连接，一旦连接建成，浏览器和服务器就可以通过套接字接口访问 TCP。HTTP 服务器通过 TCP 的周知端口 80 监听客户向它发出的连接请求。早期 HTTP 1.0 每次请求都需要建立一次 TCP 连接，服务器发回响应后 TCP 连接被释放，这是一种非持续连接。HTTP 1.1 支持持续连接，并把它作为默认选择，对于用户连续的多个访问请求，TCP 连接不会被释放，从而减少了建立连接的时延开销。

2.2.2 HTTP 报文格式

HTTP 定义的报文格式有两种：请求报文（request）和响应报文（response）。

1. 请求报文

请求报文格式如图 2-4 所示，第一行称为请求行（request line）。请求行有 3

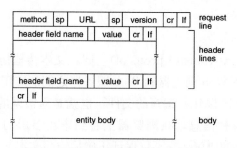

图 2-4　HTTP 请求报文的基本格式

个字段,即方法字段(method)、URL 字段(URL),这个字段可以是一个"/",以及协议版本字段(version)。HTTP 不对 URL 的长度做事先的限制,服务器必须能够处理它们服务的任何资源的 URL,并且应该能够处理无限长度的 URL。例如,报文如图 2-5 所示。

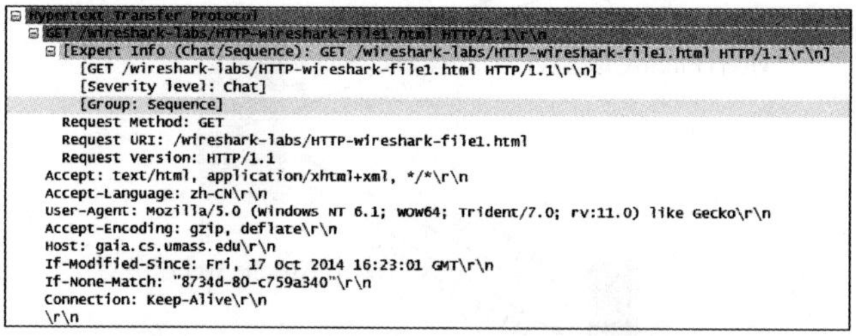

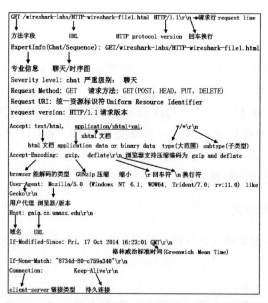

图 2-5　Wireshark 分析的 HTTP 请求报文

HTTP 具体格式内容分析如图 2-6 所示。

图2-6　Wireshark 分析的 HTTP 请求报文格式解析

常见的方法字段说明如下。
GET：请求指定的页面信息，并返回实体主体。
HEAD：只请求页面的首部。
POST：请求服务器接受所指定的文档作为对所标识的 URL 的新从属实体。
PUT：从客户端向服务器传送的数据取代指定的文档内容。
DELETE：请求服务器删除指定的页面。
OPTIONS：允许客户端查看服务器的性能。

需要特别说明的是，方法字段有若干个值可供选择，最常见的包括 GET、POST 和 HEAD。

常见的首部说明如下。
Accept-Charset：浏览器可以接受的字符集。
Accept-Encoding：浏览器可以接受的编码方案。
Accept-Language：浏览器能够接受的语言。
Host：主机和端口号。
Referer：指明被连接的目标 URL。

2. 响应报文

响应报文的基本格式如图 2-7 所示。

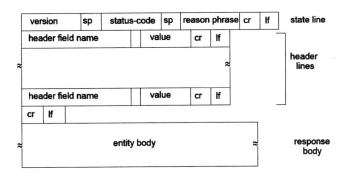

图 2-7　HTTP 响应报文的基本格式

响应报文的第一行是状态行，由协议版本（version）、状态码（status-code）及状态短语（reason phrase）组成。各部分间用空格隔开，除了最后的回车或换行，中间不允许有回车换行。

状态行中的状态码是试图理解和满足请求的三位数字的整数码，状态码用于自动控制，而状态短语是面向用户的，客户机不需要检查和显示状态短语。注意，请求报文的 HTTP 版本可以和响应报文的 HTTP 不一致。

状态码的第一位数字定义应答类型，后两位数字没有类型任务，第一位数字有如下五类值。
① 1xx 消息——接收到请求，继续进程。
② 2xx 成功——操作成功收到。
③ 3xx 重发——为了完成请求，必须采取进一步措施。
④ 4xx 客户端出错——请求包括错误的顺序或不能完成。
⑤ 5xx 服务器出错——服务器无法完成显然有效的请求。
HTTP 常见状态码的具体含义如下。
"200"：OK（成功）。
"202"：Accepted（接受）。
"203"：Non-Authoritative Information（非权威信息）。
"304"：Not Modified（条件 GET 中请求的对象未被修改）。
"404"：Not Found（请求对象未找到）。
"408"：Request Timeout（请求超时）。

实体部分：经由 HTTP 请求或应答发送的实体正文部分（如果存在）的格式与编码方式应由实体报文域决定。实体正文是通过对报文正文按某种保证安全性且便于传输的传输编码进行解码得到的。对于报文中的实体正文，其数据类型由报头中的"内容类型"与"内容编码"域决定。

实体正文＝内容编码（内容类型（数据）），"内容类型"规定了基本数据的媒体类型。

实体长度：报文的实体长度指的是在对报文进行传输编码前报文正文的长度。

2.2.3 HTTP 分析实验

针对 HTTP 的内容，通过五个实验：基本的 HTTP 请求与响应消息的交互、条件 GET、检索长文本、检索嵌入对象的页面以及 HTTP 的认证来分析协议。

在实验开始之前，请保证正确安装了 Wireshark 软件。

1. 实验一：基本的 HTTP 请求与响应消息的交互

首先利用 HTTP 下载一个简单的 HTML 文件，以此对 HTTP 进行分析，按如下步骤进行抓包。

（1）打开 Web 浏览器。

（2）启动 Wireshark 抓包软件，在窗口显示过滤器中输入"http"，以确保在抓包消息中仅显示 HTTP 消息。

（3）一分钟后，开始 Wireshark 分组捕获。

（4）在打开的 Web 浏览器下，输入 http：//gaia.cs.umass.edu/Wireshark-labs/HTTP-Wireshark-file1.html，你的浏览器将显示一个非常简单的、在线的

HTML 文件。

(5) 停止 Wireshark 分组捕获。

这时，Wireshark 如图 2-8 和图 2-9 所示。

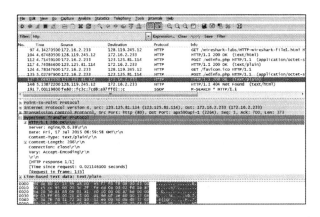

图 2-8　浏览器收到的网页在 Wireshark 中

图 2-9　浏览器收到的网页内容在 Wireshark 中

这时，浏览器显示如图 2-10 所示。

图 2-10　浏览器显示连接成功信息

HTTP 的 GET 请求报文如图 2-11 所示，可将响应报文格式与该抓取报文进行对比学习。

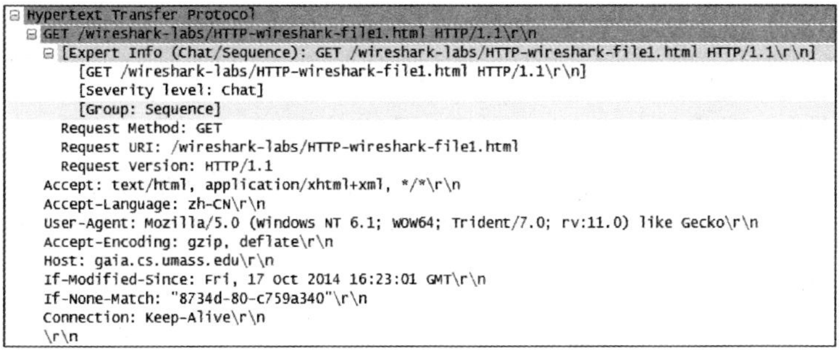

图 2-11　GET 请求报文

如图 2-12 所示，根据 Source 和 Destination 列可以确定客户机的 IP 地址为 10.2.136.28，服务器 IP 地址为 128.119.245.12。如图 2-13 所示，可以获取如下信息：HTTP 的版本是 HTTP 1.1；状态码是 200 OK，表示请求成功（报文第二行）；所使用的 Web 服务器为 Apache（报文第四行）；最后的修改时间是 Mon, 24 Sep 2012 13：22：01（报文第五行）；所包含的字节内容是 128 字节（报文第八行）；采用的是持久连接（第 10 行）。

Time	Source	Destination
6 0.887581	10.2.136.28	128.119.245.12
10 0.983133	10.2.136.28	202.117.144.2
16 0.995208	202.117.144.2	10.2.136.28
19 1.173881	128.119.245.12	10.2.136.28
23 1.543543	10.2.136.28	128.119.245.12
26 1.829214	128.119.245.12	10.2.136.28

图 2-12　HTTP IP 地址信息

```
Hypertext Transfer Protocol
 HTTP/1.1 200 OK\r\n
 Date: Mon, 24 Sep 2012 13:22:03 GMT\r\n
 Server: Apache/2.2.3 (CentOS)\r\n
 Last-Modified: Mon, 24 Sep 2012 13:22:01 GMT\r\n
 ETag: "8734d-80-7989d040"\r\n
 Accept-Ranges: bytes\r\n
 Content-Length: 128\r\n
 Keep-Alive: timeout=10, max=100\r\n
 Connection: Keep-Alive\r\n
 Content-Type: text/html; charset=UTF-8\r\n
 \r\n
```

图 2-13　HTTP 响应报文基本信息分析

2. 实验二：HTTP 的条件 GET

本实验抓包之前，需要确保清空浏览器的缓存（以 Internet Explorer 为例，选择工具→Internet 选项→删除文件）。

按如下步骤进行抓包。

（1）打开浏览器，确保浏览器的缓存清除。

（2）打开 Wireshark 抓包工具并开始抓包，在浏览器中输入 http：//gaia.cs.umass.edu/Wireshark-labs/HTTP-Wireshark-file2.html，这时浏览器将显示一个简单的文件。

（3）快速在浏览器中再次输入 URL 或进行刷新。

（4）停止 Wireshark 抓包，在过滤器中输入"http"（小写）。

接下来对比两次过程中得到的 GET 报文，第一次报文如图 2-14 所示，第二次如图 2-15 所示。

对比两次报文可以发现，在第一次的 GET 报文没有"If-Modified-Since"字段，在第二次的报文段中含有"If-Modified-Since"一行。

```
Hypertext Transfer Protocol
  GET /ethereal-labs/lab2-2.html HTTP/1.1\r\n
    Host: gaia.cs.umass.edu\r\n
    User-Agent: Mozilla/5.0 (windows; U; Windows NT 5.1; en-US; rv:1.0.2) Ge
    Accept: text/xml,application/xml,application/xhtml+xml,text/html;q=0.9,t
    Accept-Language: en-us, en;q=0.50\r\n
    Accept-Encoding: gzip, deflate, compress;q=0.9\r\n
    Accept-Charset: ISO-8859-1, utf-8;q=0.66, *;q=0.66\r\n
    Keep-Alive: 300\r\n
    Connection: keep-alive\r\n
    \r\n
    [Full request URI: http://gaia.cs.umass.edu/ethereal-labs/lab2-2.html]
```

图 2-14　第一次 GET 请求报文

```
Hypertext Transfer Protocol
  GET /ethereal-labs/lab2-2.html HTTP/1.1\r\n
    Host: gaia.cs.umass.edu\r\n
    User-Agent: Mozilla/5.0 (windows; U; Windows NT 5.1; en-US; rv:1.0.2) Ge
    Accept: text/xml,application/xml,application/xhtml+xml,text/html;q=0.9,t
    Accept-Language: en-us, en;q=0.50\r\n
    Accept-Encoding: gzip, deflate, compress;q=0.9\r\n
    Accept-Charset: ISO-8859-1, utf-8;q=0.66, *;q=0.66\r\n
    Keep-Alive: 300\r\n
    Connection: keep-alive\r\n
    If-Modified-Since: Tue, 23 Sep 2003 05:35:00 GMT\r\n
    If-None-Match: "1bfef-173-8f4ae900"\r\n
    Cache-Control: max-age=0\r\n
    \r\n
    [Full request URI: http://gaia.cs.umass.edu/ethereal-labs/lab2-2.html]
```

图 2-15　第二次 GET 请求报文

需要注意的是，对于第二次 GET 报文的响应报文会出现 304 状态码，如图 2-16 所示，状态代码为 304，状态短语为 Not Modified，并且没有返回内容，因为浏览器缓存中已经存在该文件的最新副本，不会再返回内容。

```
Hypertext Transfer Protocol
  HTTP/1.1 304 Not Modified\r\n
    Date: Tue, 23 Sep 2003 05:35:53 GMT\r\n
    Server: Apache/2.0.40 (Red Hat Linux)\r\n
    Connection: Keep-Alive\r\n
    Keep-Alive: timeout=10, max=99\r\n
    ETag: "1bfef-173-8f4ae900"\r\n
    \r\n
```

图 2-16　条件 GET 的响应报文

3. 实验三：长文档检索

本实验尝试检索一个长的文档，并分析其中 HTTP 请求和响应报文的特征。按如下步骤进行抓包。

（1）打开浏览器，确保浏览器的缓存清除。

（2）打开 Wireshark 抓包软件并开始抓包，在浏览器中输入 http：//gaia.cs.umass.edu/Wireshark-labs/HTTP-Wireshark-file3.html，这时，你的浏览器会显示一个长文档。

（3）停止 Wireshark 抓包，在过滤器中输入"http"。

报文格式如图 2-17 所示。

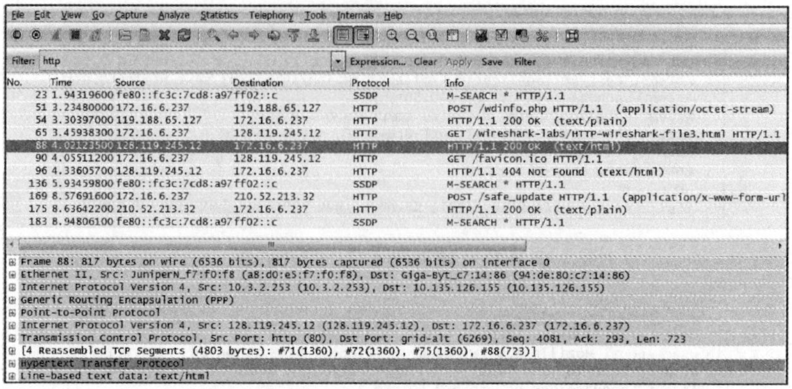

图 2-17　长文档检索报文抓包

通过报文段，如图 2-18 所示，可以看出为了运输这一个长文档构成的响应消息报文，TCP 使用了 4 个报文段，并且总长度是 4803 字节。这说明，当 HTTP 协议需要传输的消息超过一定长度时，需要多个 TCP 共同运输，原因在于 TCP 所能承载的有效载荷的大小不是无限的。

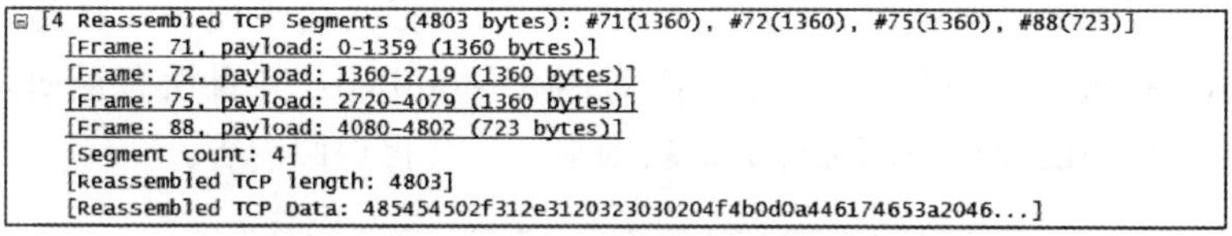

图 2-18　报文段分段显示

4. 实验四：检索包含嵌入对象的 HTML 文件

本实验分析利用 HTTP 检索包含了嵌入对象的 HTML 文件时请求和响应报文的交互，重点分析所请求对象与其页面不在同一个服务器的情形。

按如下步骤进行抓包。

（1）打开浏览器，确保浏览器的缓存清除。

（2）打开 Wireshark 抓包软件并开始抓包，在浏览器中输入 http：//gaia.cs.umass.edu/Wireshark-labs/HTTP-Wireshark-file4.html，浏览器将会展示一个简短的 HTML 文件和两张图片，这两张图片并不包含在 HTML 文件中，而图片的 URL 包含在 HTML 文件中，你的浏览器将要从指定的网站获取这些图片。

（3）停止 Wireshark 抓包，在过滤器中输入"http"。

报文信息如图 2-19 所示。

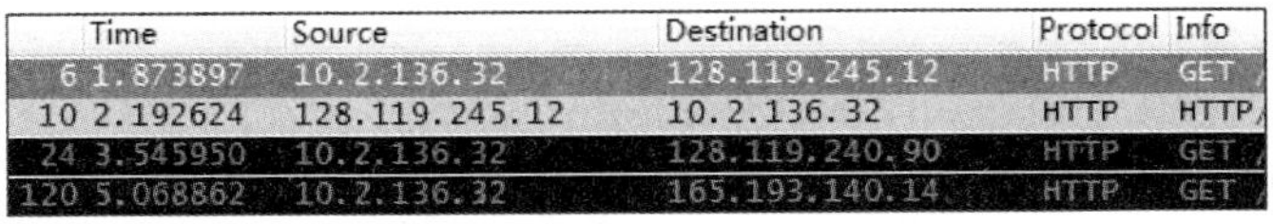

图 2-19　含有嵌入图片 HTML 文件的 HTTP 报文

如图 2-19 所示，可以看到，浏览器发送了 3 次 HTTP GET Request 消息，发送的地址分别是 128.119.245.12、128.119.240.90、165.193.140.14。如图 2-20 所示，可以看到，在第一次请求图片未得到回复时，第二次图片请求已经发出。

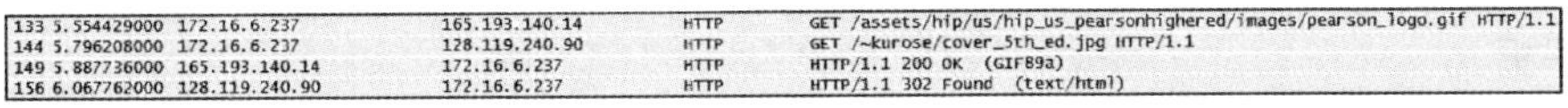

图 2-20　嵌入图片的请求和响应报文

5. 实验五：HTTP 的认证

本实验将访问一个有密码保护的网站，通过抓包访问过程来分析 HTTP 认证的特性，在这里使用网站 http：//gaia.cs.umass.edu/Wireshark-labs/protected_pages/ HTTP-Wireshark-file5.html，这是一个需要认证的网站，用户名为 Wireshark-students，密码为 network。现在，按如下步骤抓包。

（1）打开浏览器，确保浏览器的缓存清除。

（2）打开 Wireshark 抓包软件并开始抓包，在浏览器中输入 http：//gaia.cs.umass.edu/Wireshark-labs/protected_pages/ HTTP-Wireshark-file5.html，这时，填写你的用户名和密码。

（3）停止 Wireshark 抓包，在过滤器中输入"http"。抓包结果的第一个 HTTP 响应报文如图 2-21 所示。

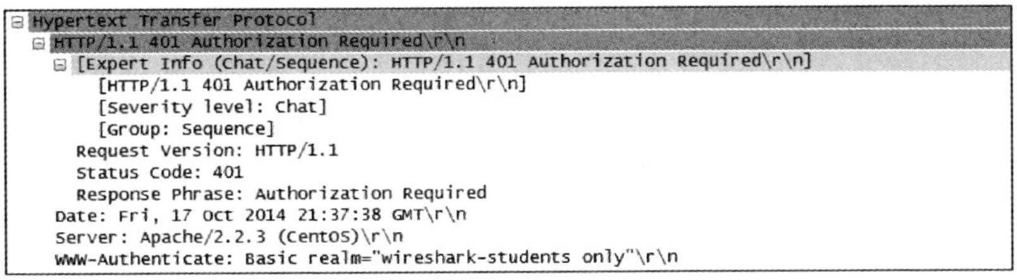

图 2-21　HTTP 认证报文

如图 2-21 所示，可以看到，返回的代码和响应是 401 Authorization Required，这说明这是个有认证保护的报文，在输入了认证信息后，第二次访问这个网站时会得到如图 2-22 所示的报文。

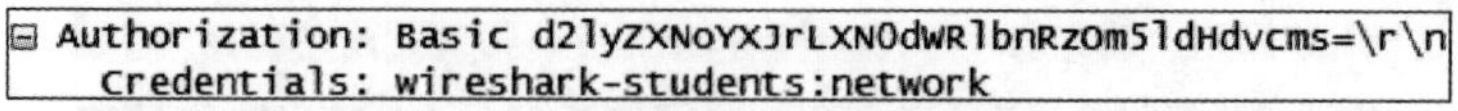

```
Authorization: Basic d2lyZXNoYXJrLXN0dWRlbnRzOm5ldHdvcms=\r\n
    Credentials: wireshark-students:network
```

图 2-22　HTTP 认证交互报文

2.3　域名系统

域名系统（domain name system，DNS）的主要功能是提供域名和 IP 地址映射关系，把容易记忆的域名转换成相应的 IP 地址。一般人在上网时，并不习惯直接输入 IP 地址，而是输入域名，这时就需要通过 DNS 服务器来帮助完成域名解析。

2.3.1　DNS 涉及的基本概念

1. 域名

域名（domain name）通常是用户所在的主机名。域名格式是由若干部分组成的，每个部分又称子域名，它们之间用"."分开，每个部分最少由两个字母或数字组成。域名通常按分层结构来构造，每个子域名都有其特定的含义。从右到左，子域名分别表示不同的国家或地区的名称（只有美国可以省略表示国家的顶级域名）、组织类型、组织名称、分组织名称和计算机名称等，如 www.jwc.snnu.edu.cn 就是一个域名的典型例子。

2. 域名系统的构成

域名系统采用的是 C/S 模式，由三部分构成：域名数据库、域名服务器和地址解析器。

（1）地址解析器是客户方，负责查询域名服务器、解释从服务器返回来的应答、将信息返回给请求方等工作。

（2）域名服务器是服务器方，存储并管理着所管辖区域的域名数据库，负责接收来自地址解析器的请求，按照请求类型，进行递归与迭代查询。同时将查询结果返回给地址解析器（网络中的每台主机既可作为客户方，也可作为服务器方）。

（3）域名数据库是一个大型的分布在整个网络上的分布式数据库，存储了按层次管理的相关的数据，该层次结构可理解为一棵倒放的树，树中的每个节点均代表一个域，并存储着与该域相关的区域和资源记录（RRs），以供域名服务器查询使用。

3. 域名解析过程

当某个主机请求获得域名 www.cs.umass.edu 的 IP 地址时,其地址解析程序首先从根域名服务器获得域.edu 的域名信息,然后从.edu 的域名服务器获得其管辖下的子区域 umass.edu 的域名信息,这样逐级向下进行查询,直到到达 cs.edu.cn 子区域的域名服务器,并从中获得其相应的 IP 地址,反馈给地址解析程序,如图 2-23 所示。

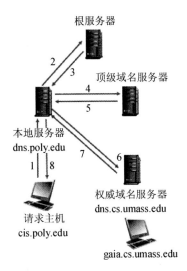

图 2-23　各种 DNS 服务的交互

综上所述,一个单一的域名服务器无法对每一个域名查询进行完整的回答,但是它可以对查询路径作出准确的响应。即当一个域名服务器存储了一个域名查询所请求域的所有授权信息,它可直接给出需要的查询结果,否则,它必须给出具有所需信息的最近的域名服务器,以便继续查询,上述过程一直持续到请求方得到正确的结果或访问某个域名服务器时出现错误。

2.3.2　递归查询和迭代查询

1. 递归查询

所谓递归查询是指接收请求的第一个域名服务器必须自始至终对请求进行处理,或向其他域名服务器进行请求且最终获得解析数据,并对请求进行应答,如图 2-24 所示。

2. 迭代查询

所谓迭代查询是指接收请求的第一个域名服务器可以返回可靠数据,也可以

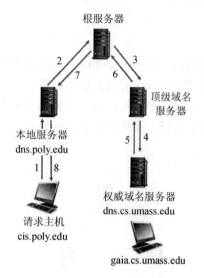

图 2-24　DNS 中的递归查询

返回指向其他服务器的指针（相当于将查询的接力棒传给最接近的域名服务器）。非递归查询方式与递归查询方式相比响应速度快。

2.3.3　DNS 资源记录

实现 DNS 分布式数据库的所有 DNS 服务器共同存储着资源记录（resource record，RR），RR 提供了主机名和 IP 地址等关系的映射，资源记录是一个包含了下列字段的 4 元组，即

(Name，Value，Type，TTL)

TTL 是该记录的生存时间，它决定了资源记录应当从缓存中删除的时间。简单地说，TTL 就是一条域名解析记录在 DNS 服务器中的存留时间。在获得记录之后，记录会在 DNS 服务器中保存一段时间，这段时间内如果再接到这个域名的解析请求，DNS 服务器将不再向服务器发出请求，而是直接返回刚才获得的记录，而这个记录在 DNS 服务器上保留的时间就是 TTL 值。

其余三个字段基于 Type 的不同类型而不同。下面介绍 Type 的四种类型。

1) Type=A

A（address）记录是用来指定主机名（或域名）对应的 IP 地址记录。用户可以将该域名下的网站服务器指向自己的 Web server 上。同时也可以设置域名的子域名。通俗来说，A 记录就是服务器的 IP，域名绑定 A 记录就是告诉 DNS，当你输入域名的时候给你引导向设置在 DNS 的 A 记录所对应的服务器。简单地说，A 记录是指定域名对应的 IP 地址。此时，Name 字段为域名，Value 字段为 IP 地址。

2) Type=NS

NS（name server）记录是域名服务器记录，用来指定该域名由哪个 DNS 服务器来进行解析。注册域名时，总有默认的 DNS 服务器，每个注册的域名都是由一个 DNS 域名服务器来进行解析的，DNS 服务器 NS 记录地址一般以以下的形式出现：ns1.domain.com 以及 ns2.domain.com 等。简单地说，NS 记录是指定由哪个 DNS 服务器解析你的域名。此时，Name 字段为域名，Value 字段为该域名所对应的权威域名服务器。

3) Type=MX

MX（mail exchanger）记录是邮件交换记录，它指向一个邮件服务器，用于电子邮件系统发邮件时，根据收信人的地址后缀来定位邮件服务器。例如，当 Internet 上的某用户要发一封信给 user@mydomain.com 时，该用户的邮件系统通过 DNS 查找 mydomain.com 这个域名的 MX 记录，如果 MX 记录存在，用户计算机就将邮件发送到 MX 记录所指定的邮件服务器上。此时，Name 字段为@符号后的域名，Value 字段为该域名所对应的邮件服务器。

4) Type=CNAME

CNAME（canonical name）别名记录，允许将多个名字映射到同一台计算机。通常用于同时提供 WWW 和 Mail 服务的计算机。例如，有一台名为"host.mydomain.com"（A 记录）的计算机，它同时提供 WWW 和 Mail 服务，为了便于用户访问服务，可以为该计算机设置两个别名（CNAME），即 WWW 和 Mail，这两个别名的全称就是"www.mydomain.com"和"mail.mydomain.com"，实际上它们都指向"host.mydomain.com"。此时，Name 字段为域名别名，Value 字段为该域名所对应的规范名。

2.3.4 DNS 报文

DNS 只有查询和回答报文，而且这两种报文共享相同的格式，如图 2-25 所示。

标识符（2 字节）：这个字段可以看成 DNS 报文的 ID，对于相关联的请求报文和应答报文，这个字段是相同的，由此可以区分 DNS 应答报文是哪个请求报文的响应。

标志(2 字节)：对该部分进行逐比特分析，如图 2-26 所示。

QR(1 比特)：查询/响应的标志位，1 为响应，0 为查询。

opcode(4 比特)：定义查询或响应的类型（若为 0 则表示是标准的，若为 1 则是反向的，若为 2 则是服务器状态请求）。

图 2-25　DNS 报文格式

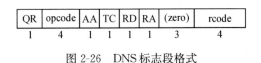

图 2-26　DNS 标志段格式

AA（1 比特）：授权回答的标志位，在响应报文中有效，1 表示名字服务器是权限服务。

TC（1 比特）：截断标志位。1 表示响应已超过 512 字节并已被截断。

RD（1 比特）：该位为 1 表示客户端希望得到递归回答。

RA（1 比特）：只能在响应报文中置为 1，表示可以得到递归响应。

(zero)（3 比特）：都是 0，保留字段。

rcode（4 比特）：返回码一般有 7 个，表示响应的差错状态，通常为 0 和 3，含义为

　　　　　　　　0　　　无差错
　　　　　　　　1　　　格式差错
　　　　　　　　2　　　问题在域名服务器上
　　　　　　　　3　　　域参照问题

下面对问题数、资源记录数、授权资源记录数和额外资源记录数进行介绍，这四个字段都是 2 字节，分别对应下面的查询问题、回答、授权和额外信息部分的数量。一般问题数都为 1，DNS 查询报文中，资源记录数、授权资源记录数和额外资源记录数都为 0。

在正文部分，查询问题部分格式如图 2-27 所示。

图 2-27 DNS 查询字段基本格式

查询名部分长度不定,一般为要查询的域名(也会有 IP 的时候,即反向查询)。此部分由一个或者多个标识符序列组成,每个标识符以首字节数的计数值来说明该标识符长度,每个名字以 0 结束,计数字节数必须在 0~63 之间,该字段无需填充字节。

回答字段、授权字段和附加信息字段均采用资源记录(resource record, RR)的相同格式,如图 2-28 所示。

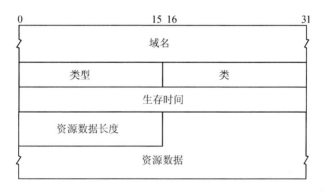

图 2-28 DNS 回答字段基本格式

域名字段(不定长或 2 字节):记录中资源数据对应的名字,它的格式和查询名字段格式相同。当报文中域名重复出现时,就需要使用 2 字节的偏移指针来替换。

类型(2 字节)、类(2 字节):含义与查询问题部分的类型和类相同。

生存时间(4 字节):该字段表示资源记录的生命周期(以秒为单位),一般用于当地址解析程序取出资源记录后决定保存及使用缓存数据的时间。

资源数据长度(2 字节):表示资源数据的长度(以字节为单位,如果资源数据为 IP,则为 100(4))。

资源数据:该字段是可变长字段,表示按查询段要求返回的相关资源记录的数据。

Wireshark 下 DNS 报文如图 2-29 所示。

```
■ Domain Name System (response)
    [Request In: 739]
    [Time: 0.003302000 seconds]
    Transaction ID: 0x8759
  ⊞ Flags: 0x8180 (Standard query response, No error)
    Questions: 1
    Answer RRs: 1
    Authority RRs: 2
    Additional RRs: 0
  ⊟ Queries
    ⊞ reg.eol.cn: type A, class IN
  ⊟ Answers
    ⊟ reg.eol.cn: type A, class IN, addr 121.194.3.182
        Name: reg.eol.cn
        Type: A (Host address)
        Class: IN (0x0001)
        Time to live: 5 minutes
        Data length: 4
        Addr: 121.194.3.182
  ⊟ Authoritative nameservers
    ⊟ eol.cn: type NS, class IN, ns dns1.cernet.cn
        Name: eol.cn
        Type: NS (Authoritative name server)
        Class: IN (0x0001)
        Time to live: 42 minutes, 20 seconds
        Data length: 14
        Name server: dns1.cernet.cn
    ⊞ eol.cn: type NS, class IN, ns dns2.cernet.cn
```

图 2-29　Wireshark 显示的 DNS 报文

Wireshark 下 DNS 报文数据格式如图 2-30 所示。

```
0020  a1 a5 00 35 e6 e4 00 61 77 56 87 59 81 80 00 01    ...5...a wV.Y....
0030  00 01 00 02 00 00 03 72 65 67 03 65 6f 6c 02 63    .......r eg.eol.c
0040  6e 00 00 01 00 01 c0 0c 00 01 00 01 00 00 01 2c    n........ .......,
0050  00 04 79 c2 03 b6 c0 10 00 02 00 01 00 00 09 b6    ..y..... ........
0060  00 0e 04 64 6e 73 31 06 63 65 72 6e 65 74 c0 14    ...dns1. cernet..
0070  c0 10 00 02 00 01 00 00 09 ec 00 07 04 64 6e 73    ........ .....dns
0080  32 c0 3d                                           2.=
```

图 2-30　Wireshark 显示的 DNS 协议数据

2.3.5　DNS 协议分析实验

在本实验中，需要学会使用 nslookup 工具查询并分析 Internet 域名信息、DNS 服务器。学会使用 ipconfig 工具进行分析，会用 Wireshark 分析 DNS 报文，对 DNS 协议进行全面的学习与了解。

实验之前，请确保计算机已经连上了 Internet 并成功安装 Wireshark 软件。

1. 实验一：nslookup 的使用

nslookup 是命令提示符命令，可以对指定的 DNS 服务器进行查询，查询的 DNS 服务器可以是一个顶级域名服务器、根域名服务器、权威域名服务器。为了完成这项任务，nslookup 发送一个 DNS 查询到指定的 DNS 服务器，相应的 DNS 收到后回复并显示结果。

一般来说，nslookup 可以运行一个、两个或两个以上的选项。DNS 服务器是可选参数，如果没有明确指定，将发送查询到本地 DNS 服务器。

nslookup 的基本语法为

　　　　nslookup - option1 - option2 host-to-find dns-server

按如下步骤进行抓包。

基于 Windows 平台时,打开"命令提示符",在命令行下可运行 nslookup。

如果想知道一个在亚洲的 Web 服务器的 IP 地址,如百度首页,可以输入"nslookup www.baidu.com",用本机默认的 DNS 服务器对百度网页域名进行解析,如图 2-31 所示,得到了一个 IP 地址,通过在浏览器地址栏输入该 IP 地址时,却发现了这个地址无法找到百度主页。于是为了得到一个可靠的答案,可使用国内几个较为常用的 DNS 服务器,这里使用的是 114.114.114.114 作为 DNS 服务器,这是一个较为通用的 DNS 服务器,当然也存在一些其他可靠的 DNS 服务器,如 8.8.8.8 等。在图 2-32 中,可以看出利用 114.114.114.114 找出了百度的一个域名为 www.a.shifen.com,域名服务器地址为 180.97.33.108,这时就可以通过这个 IP 地址来访问百度地址。对比图 2-33,可以看到当利用 8.8.8.8 解析后得到了与之前相同的百度域名,但是域名服务器的地址是不同的,但是毋庸置疑的是,这些地址都是可以正确访问百度网站的。因此,一个网页可以有多个域名,每个域名可以有多个 IP 地址。

图 2-31　nslookup 利用默认 DNS 服务器查询百度网址

图 2-32　常用 DNS 服务器 114.114.114.114 查询百度网址

图 2-33　常用 DNS 服务器 8.8.8.8 查询百度网址

这时，可以获得百度地址是 180.97.33.108 或 61.135.169.121，别名是 www.a.shifen.com。还有另外几种类型格式进行 DNS 地址解析，如图 2-34 所示，请求的 DNS 服务器是 dns-xgx-man.169cnc.net，IP 地址是 221.11.1.67，得到 oxford.edu 的权威域名有两个，分别为 dns2.ox.ac.uk 和 dns0.ox.ac.uk。

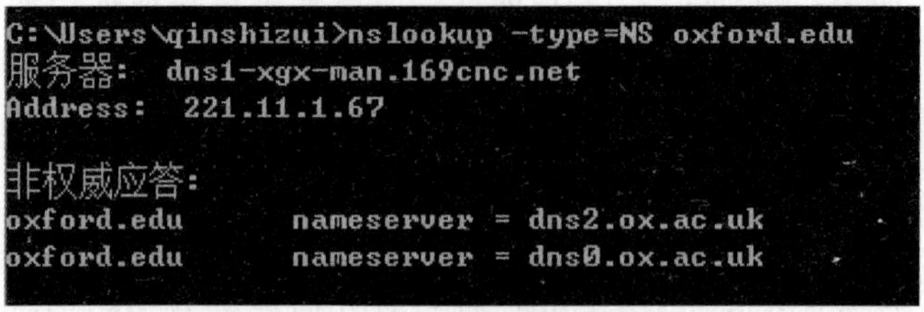

图 2-34　NS 类型访问

2. 实验二：ipconfig 工具的使用

ipconfig（Windows 平台）以及 ifconfig（Linux/Unix/Mac 平台）都是常用的网络调试工具，在这里只介绍 ipconfig，ifconfig 的使用方法类似。

运行如下命令，并观察和分析结果。

1）ipconfig /all

输入之后，你将看到如图 2-35 所示的结果，在图中可以获得主机地址记录等信息。

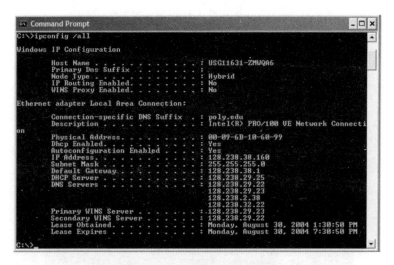

图 2-35　ipconfig 命令执行结果

2）ipconfig /displaydns

显示主机的 DNS 缓存记录信息，包括每个条目剩余生存时间（TTL）等。

3）ipconfig /flushdns

清除主机的 DNS 缓存。

3. 实验三：用 Wireshark 跟踪 DNS

（1）使用 ipconfig /flushdns 清空 DNS 缓存。

（2）打开你的浏览器，清空你的浏览器缓存（如使用 Internet Explorer，转到工具菜单并选择 Internet 选项，然后在"常规"选项卡上选择"删除文件"）。

（3）打开 Wireshark 并输入"ip.addr == your_ip_address"进入过滤器，可用 ipconfig 获得本机的 IP 地址。

（4）开始在 Wireshark 数据包捕获。

（5）打开浏览器，访问网页 http://www.ietf.org。

（6）停止捕获数据包。

这时，可以看到两则报文，如图 2-36 和图 2-37 所示。

请求消息如图 2-36 所示。

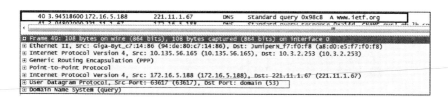

图 2-36　Wireshark 显示 DNS 请求消息

响应消息如图 2-37 所示。

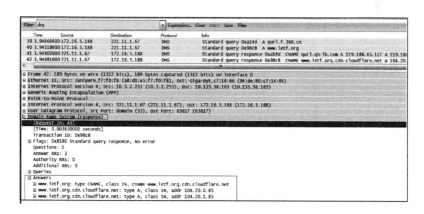

图 2-37　Wireshark 显示 DNS 响应消息

可以看到，请求消息和响应消息的 IP 地址是对应的，请求消息的源 IP 地址是相应消息的目的 IP 地址，即 172.16.5.188；所用的端口号也呈对应关系，请求消息的源端口号和响应消息的目的端口号一致,为 63617,请求消息的目的端口号和响应消息的源端口号一致。在传输中,可以看到,使用的是 UDP 协议传输。

4. 实验四:执行命令 nslookup www.mit.edu

接下来,使用 nslookup 进行相关操作并抓包分析,按如下步骤进行。
(1) 打开 Wireshark 并开始抓包。
(2) 执行命令 nslookup www.mit.edu。
(3) 停止捕获数据包。
这时可以看到如图 2-38 所示的执行结果分析。

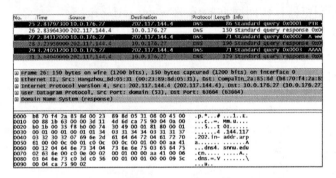

图 2-38 nslookup www.mit.edu 执行结果分析

首先,DNS 的查询报文源端口号是 53,目的 IP 地址是 10.0.176.27,源 IP 地址是 202.117.144.4,这个地址和本地服务器的地址是一致的。

在这个消息中包含了 3 个答案,如图 2-39 和图 2-40 所示,第一个和第二个包含名字、类型、生命周期、数据长度、别名,最后一个包含名字、类型、生命周期、数据长度、地址。

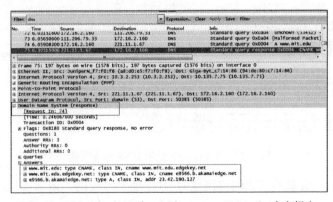

图 2-39 Wireshark 显示 nslookup www.mit.edu 响应报文

```
Answers
  www.mit.edu: type CNAME, class IN, cname www.mit.edu.edgekey.net
    Name: www.mit.edu
    Type: CNAME (Canonical NAME for an alias) (5)
    Class: IN (0x0001)
    Time to live: 1800
    Data length: 25
    CNAME: www.mit.edu.edgekey.net
  www.mit.edu.edgekey.net: type CNAME, class IN, cname e9566.b.akamaiedge.net
    Name: www.mit.edu.edgekey.net
    Type: CNAME (Canonical NAME for an alias) (5)
    Class: IN (0x0001)
    Time to live: 38
    Data length: 21
    CNAME: e9566.b.akamaiedge.net
  e9566.b.akamaiedge.net: type A, class IN, addr 23.42.190.127
    Name: e9566.b.akamaiedge.net
    Type: A (Host Address) (1)
    Class: IN (0x0001)
    Time to live: 20
    Data length: 4
    Address: 23.42.190.127 (23.42.190.127)
```

图 2-40　Wireshark 显示报文具体应答

5. 实验五：执行命令 nslookup-type=NS mit.edu

其他内容不变，执行命令 nslookup-type=NS mit.edu，结果如图 2-41 所示。

图 2-41　Wireshark 显示 nslookup-type=NS mit.edu 执行结果

在发送这个消息查询后，Type（类型）是 NS，如图 2-42 所示，且请求的消息不包含答案。

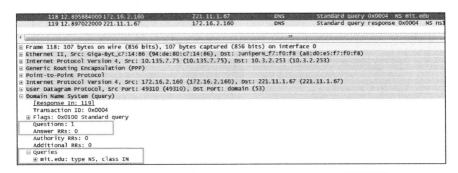

图 2-42　Wireshark 显示 nslookup-type=NS mit.edu 请求报文

如图 2-43 所示，nslookup-type＝NS mit.edu 的响应报文中，可以看到响应消息提供了 8 个域名服务器回答，但它们是非权威的答案。它不提供域名服务器的 IP 地址。

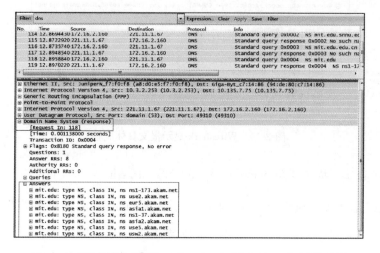

图 2-43　Wireshark 显示 nslookup-type＝NS mit.edu 响应报文

6. 实验六：执行命令 nslookup www.aiit.or.kr bitsy.mit.edu

在这个报文段中，如图 2-44 所示，DNS 的查询报文源端口号是 53，目的 IP 地址是 10.0.176.27，源 IP 地址是 18.72.0.3，很明显这个地址和本地服务器的地址不是一致的。

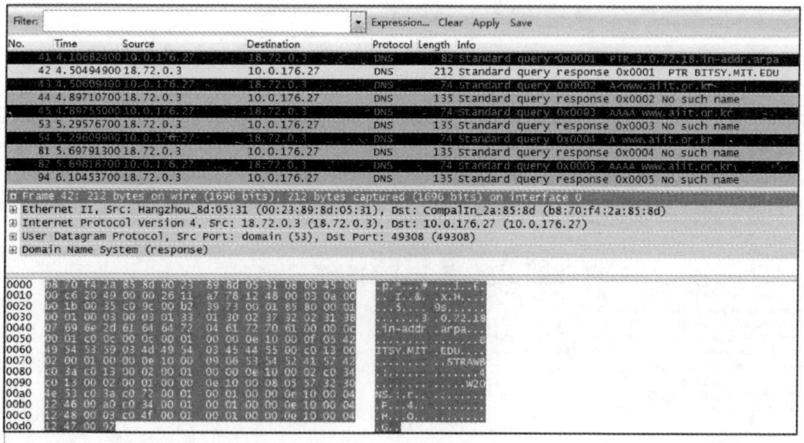

图 2-44　Wireshark 显示 nslookup www.aiit.or.kr bitsy.mit.edu 抓包内容

在发送这个消息查询后，Type（类型）是 A，如图 2-45 所示，并且请求消息不包含答案。

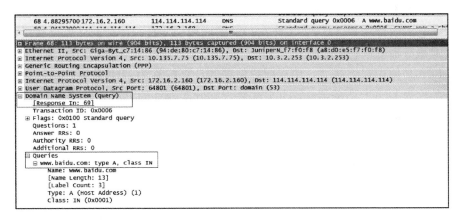

图 2-45　Wireshark 显示 nslookup www.aiit.or.kr bitsy.mit.edu 查询报文

如图 2-46 所示，给出了递归查询的简单介绍，通过实验分析可以自行分析递归查询与迭代查询之间的异同。

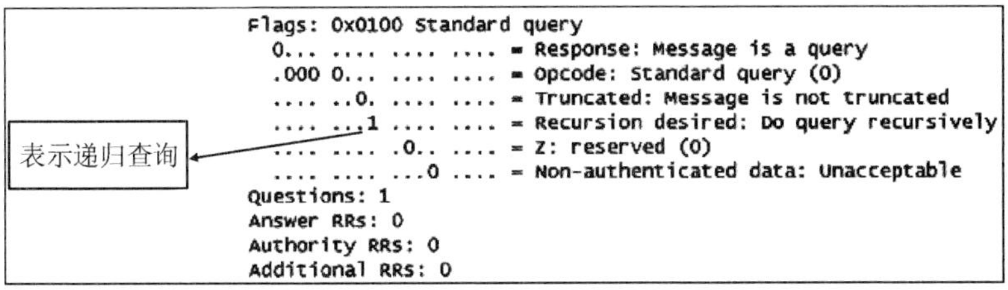

图 2-46　DNS 的递归查询报文格式

2.4　动态主机配置协议

接入 Internet 的主机必须具有一个 IP 地址，IP 地址一般是人工事先分配好的，它适用于相对静态的环境。因此，当计算机可能改变在网络上的位置时，需要改变配置参数，而且即使是相对静态的环境，对于一个大规模的网络，人工管理和分配大量的 IP 地址也容易出现重复分配等错误。同时，注意到如果某些计算机分配了 IP 地址后存在较长时间不使用而造成资源浪费。因此，可以考虑采用动态主机配置协议（dynamic host configuration protocol，DHCP）来动态分配 IP 地址。

DHCP 提供了动态配置 IP 地址的机制。除此之外，还可以提供子网掩码、默认路由器 IP 地址和域名服务器 IP 地址等配置信息，实现即插即用。

2.4.1 DHCP 简介

DHCP 全称是 dynamic host configuration protocol，中文名为动态主机配置协议，它的前身是 BOOTP，工作在 OSI 的应用层，是一种帮助计算机从指定的 DHCP 服务器获取它们的配置信息的自举协议。

DHCP 使用 C/S 模式，请求配置信息的计算机称为 DHCP 客户端，而提供信息的称为 DHCP 服务器。DHCP 为客户端分配地址的方法有三种：手工配置、自动配置、动态配置。DHCP 最重要的功能就是动态分配。除了 IP 地址，DHCP 分组还为客户端提供其他的配置信息，如子网掩码。这使得客户端无需用户动手就能自动配置连接网络。

2.4.2 DHCP 的工作流程

DHCP 的协议交互大体分为四个阶段：发现阶段、提供阶段、请求阶段和确认阶段。

（1）发现阶段，即 DHCP 客户机寻找 DHCP 服务器的阶段。DHCP 客户机以广播方式（因为 DHCP 服务器的 IP 地址对于客户机是未知的）发送 DHCP Discover 发现信息来寻找 DHCP 服务器，即向地址 255.255.255.255 发送特定的广播信息。网络上每一台安装了 TCP/IP 协议的主机都会接收到这种广播信息，但只有 DHCP 服务器才会做出响应。

（2）提供阶段，即 DHCP 服务器提供 IP 地址的阶段。在网络中接收到 DHCP Discover 发现信息的 DHCP 服务器都会做出响应，它从尚未出租的 IP 地址中挑选一个分配给 DHCP 客户机，向 DHCP 客户机发送一个包含出租的 IP 地址和其他设置的 DHCP Offer 提供信息。

（3）请求阶段，即 DHCP 客户机选择某台 DHCP 服务器提供的 IP 地址的阶段。如果有多台 DHCP 服务器向 DHCP 客户机发来的 DHCP Offer 提供信息，则 DHCP 客户机只接受第一个收到的 DHCP Offer 提供信息，接着就以广播方式回答一个 DHCP Request 请求信息，该信息中包含向它所选定的 DHCP 服务器请求 IP 地址的内容。之所以要以广播方式回答，是为了通知所有的 DHCP 服务器，他将选择某台 DHCP 服务器所提供的 IP 地址。

（4）确认阶段，即 DHCP 服务器确认所提供的 IP 地址的阶段。当 DHCP 服务器收到 DHCP 客户机回答的 DHCP Request 请求信息之后，它便向 DHCP 客户机发送一个包含它所提供的 IP 地址和 DHCP ACK 确认信息，告诉 DHCP 客户机可以使用它所提供的 IP 地址，然后 DHCP 客户机便将其 TCP/IP 协议与网卡绑定。另外，除 DHCP 客户机选中的服务器，其他的 DHCP 服务器都将收回曾提供的 IP 地址。

发现-提供-请求-确认，是动态分配 IP 地址的四个基本步骤，除此之外，DHCP 还具有以下几个阶段。

（1）重新登录。确认后的 DHCP 客户机重新登录网络时，就不需要再发送 DHCP Discover 发现信息，而是直接发送包含前一次所分配的 IP 地址的 DHCP Request 请求信息。当 DHCP 服务器收到这一信息后，它会尝试让 DHCP 客户机继续使用原来的 IP 地址，并回答一个 DHCP ACK 确认信息。如果此 IP 地址已无法再分配给原来的 DHCP 客户机使用时（如此 IP 地址已分配给其他 DHCP 客户机使用），DHCP 服务器给 DHCP 客户机回答一个 DHCP NACK 否认信息。当原来的 DHCP 客户机收到此 DHCP NACK 否认信息后，它就必须重新发送 DHCP Discover 发现信息来请求新的 IP 地址。

（2）更新租约。DHCP 服务器向 DHCP 客户机出租的 IP 地址一般都有一个租借期限，期满后 DHCP 服务器便会收回出租的 IP 地址。如果 DHCP 客户机要延长其 IP 租约，则必须更新其 IP 租约。DHCP 客户机启动时和 IP 租约期限过一半时，DHCP 客户机都会自动向 DHCP 服务器发送更新其 IP 租约的信息。

2.4.3 DHCP 的报文格式

DHCP 的报文格式如图 2-47 所示。

OP(1)	Htype(1)	Hlen(a)	Hops(1)
Transaction ID(4)			
Seconds(2)		Flags(2)	
Ciaddr(4)			
Yiaddr(4)			
Siaddr(4)			
Giaddr(4)			
Chaddr(16)			
Sname(64)			
File(128)			
Options(variable)			

图 2-47 DHCP 报文基本格式

OP：若是 client 送给 server 的分组，置为 1，反向为 2。

Htype：硬件类别，Ethernet 为 1。

Hlen：硬件长度，Ethernet 为 6。

Hops：若数据包需经过 router 传送，每站加 1，若在同一网内，为 0。

Transaction ID：事务 ID，是个随机数，用于客户和服务器之间匹配请求和相应消息。

Seconds：由用户指定的时间，指开始地址获取和更新进行后的时间。

Flags：从 0～15bit，最左一位为 1 时表示 server 将以广播方式传送封包给 client。

Ciaddr：用户 IP 地址。

Yiaddr：客户 IP 地址。

Siaddr：用于 bootstrap 过程中的 IP 地址。

Giaddr：转发代理（网关）IP 地址。

Chaddr：client 的硬件地址。

Sname：可选 server 的名称，以 0x00 结尾。

File：启动文件名。

Options：厂商标识，可选的参数字段。

2.4.4 DHCP 分析实验

在 DHCP 实验中，需要理解 DHCP 的工作原理，了解 DHCP 分配 IP 地址的过程并掌握 DHCP 服务器的设置。

在实验开始之前，请保证计算机可以正常联网，并正确安装 Wireshark 软件。

按如下步骤进行抓包。

（1）打开命令提示窗口，输入"ipconfig /release"。该命令将释放你的 IP 地址，此时你的主机的 IP 地址变为 0.0.0.0。

（2）启动 Wireshark 并开始抓包。

（3）现在回到 Windows 命令提示符下输入"ipconfig /renew"，这将指示你的主机获取网络配置，包括一个新的 IP 地址。

（4）等到"ipconfig /renew"执行终止，再次输入"ipconfig /renew"。

（5）当第二次"ipconfig /renew"终止，输入命令"ipconfig /release"释放先前分配的 IP 地址。

（6）最后，再次输入"ipconfig /renew"为你的计算机分配一个 IP 地址。

（7）停止抓包。

命令窗口显示的内容如图 2-48 所示。

第 2 章 应用层典型协议分析

图 2-48 Command 命令显示 ipconfig/renew

（8）接下来在协议过滤框中输入"bootp"，回车，这时只会看到 Wireshark 报文。如图 2-49 所示，看到主机获得的 IP 地址为 192.168.1.108。下面通过实验报文来进入协议分析。

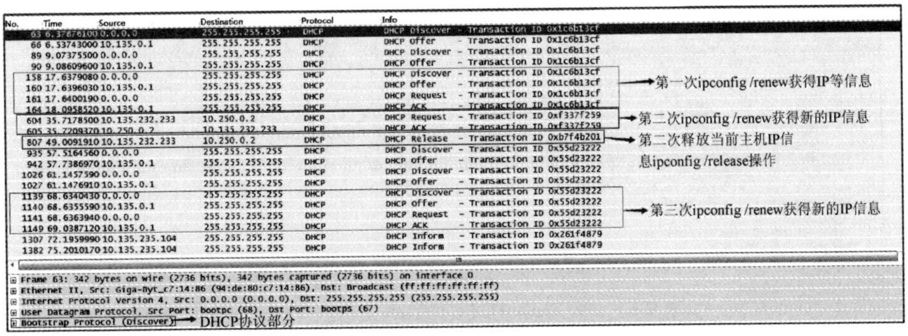

图 2-49 DHCP 四种消息报文

注意到，DHCP 有四种报文类型：DHCP Discover、DHCP Offer、DHCP Request、DHCP ACK。针对于四种类型的特点进行具体分析。

先对四个整体进行分析，如图 2-50 所示，可以看出它们都是采用 UDP 协议进行传输的。

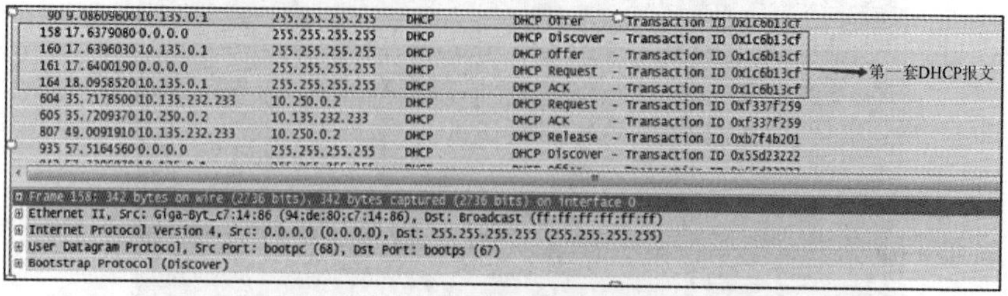

图 2-50　Wireshark 显示 DHCP 第一组报文信息

在实验中，根据数据分析出了四个报文的源端口号和目的端口号，如表 2-1 所示，可看出发现和请求的端口号是相同的，提供和确认的端口号也是相同的。

表 2-1　DHCP 四种类型报文端口号

	Discover	Offer	Request	ACK
Frame	158	160	161	164
Src port	68	67	68	67
Des port	67	68	67	68

并且如图 2-51 显示，看到数据链路层的源地址是 94:de:80:c7:14:86，这是由计算机自身在出厂时所决定的，与网络无关，目的地址是 ff:ff:ff:ff:ff:ff。对比图 2-52 和图 2-53，还可以注意到，在第一批报文内部的 Transaction-ID 是相同的，是 0xf337f259（4080530009），这个代表着事务的序列号，交易 ID 字段是一个 32 位的二进制随机数，由客户端创建，它是用来沟通请求和响应消息的 DHCP 客户端和服务器。

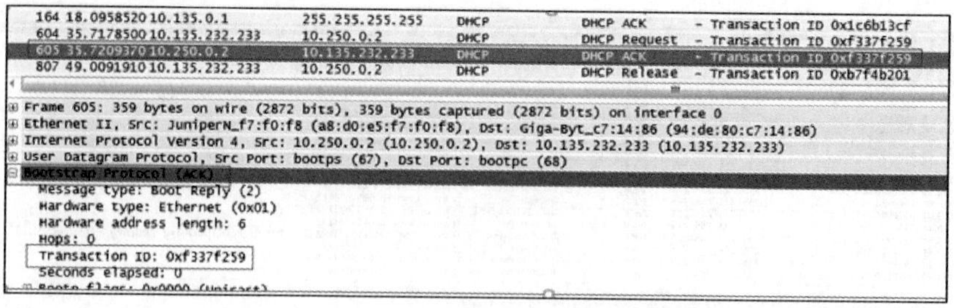

图 2-51　Wireshark 显示 DHCP 第一批报文

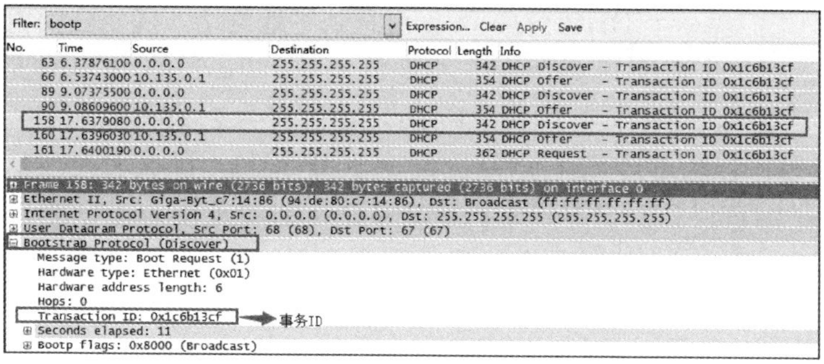

第 2 章 应用层典型协议分析

图 2-52　Wireshark 显示 DHCP Discover 消息

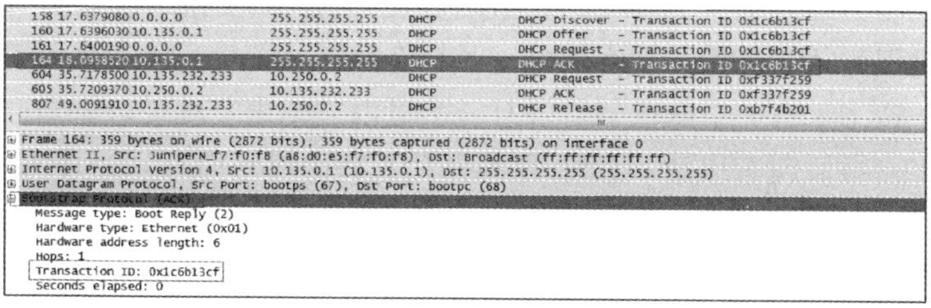

图 2-53　Wireshark 显示 DHCP ACK 消息

同时，对于三批报文，它们的 Transaction ID 是不同的，详见图 2-54。

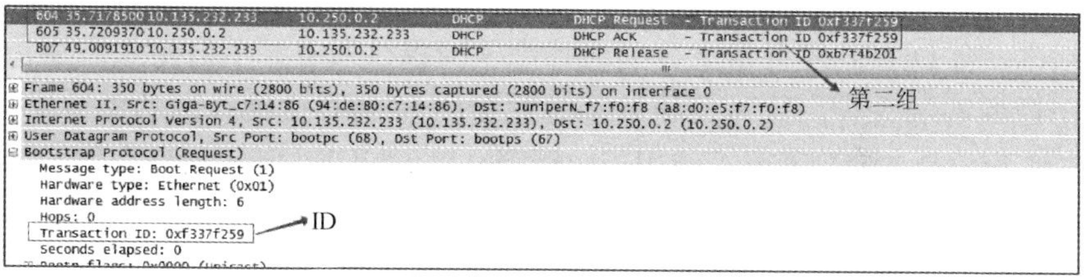

图 2-54　Wireshark 显示 DHCP 报文间 Transaction ID 分析

主机通过 DHCP 为网络客户机分配动态的 IP 地址，在这四个报文中地址是如何变换的呢，通过报文段中源 IP 地址及目的 IP 地址的分析，可以发现表 2-2 展现的答案。

表 2-2　四种报文的地址分析

	Discover	Offer	Request	ACK
Frame	158	160	161	164
Src IP	0.0.0.0	10.135.0.1	0.0.0.0	10.135.0.1
Dst IP	255.255.255.255	255.255.255.255	255.255.255.255	255.255.255.255

通过图 2-54，可以回忆起 DHCP 的协议过程如下。首先，广播 DHCP Discover，客户端广播一个 DHCP Discover 消息到本地网络，确定任何可用的 DHCP 服务器，因此此时源 IP 地址是主机地址 0.0.0.0，目的 IP 地址是广播地址 255.255.255.255，接着响应 DHCP Offer，如果有一个 DHCP 服务器连接到本地网络，并且可以为 DHCP 客户端分配一个 IP 地址，那么它会发送一个单播 DHCP Offer 消息到 DHCP 客户端。这时的源 IP 地址就变为图中的 10.135.0.1，接着响应 DHC Request，于是主机请求包含在 DHCP Offer 消息中的 IP 地址，因此源 IP 地址为主机地址，最后 DHCP ACK 确认，如果 DHCP 客户端请求的 IP 地址仍然有效，DHCP 服务器使用 DHCP ACK 确认消息应答，现在客户端可以使用这个 IP 地址了。通过这四步以后，IP 地址就分配结束了。很多教材中将这四个过程分为寻找服务器-提供 IP 租用地址-接收租约-租约确认。这个过程比较形象，可以想象一开始你没有地址，于是你自己（主机）将自己封包，向全网索要地址，接下来有一个服务器告诉有一个可以"租用"地址，于是它张贴一个消息告诉大家可以向它申请，当你了解到了这个全网的消息，并提出你的申请，在对方收到这个消息以后，愿意你"租用"这个地址，于是向全网返回 ACK，并将地址分配给你。必须注意的是，如果在这其中的任何一步没有完成，那么客户机的 IP 地址将不被分配，仍然为 0.0.0.0，目的 IP 地址依旧是 255.255.255.255。

但随着局域网的逐步扩大，一个网络通常要被划分为多个不同的子网来实现不同子网的特殊管理要求，而虚拟局域网是分隔广播域的，如果在各个不同子网分别创建一台 DHCP 服务器来分别为每个子网提供服务，不但操作复杂，而且不利于局域网的管理。这时分配地址时就利用 DHCP 中继代理功能解决问题，那么在实验中是如何判断是否使用了中继呢？以 DHCP Offer 消息为例，给出图 2-55 所示的一个信息，在这个图中是不存在中继代理器的，因为可以看到代理

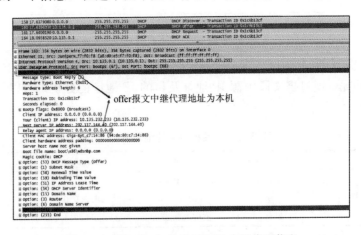

图 2-55　Wireshark 显示 DHCP 中继代理信息

IP 地址就是主机的地址，所以没有使用其他的服务器，反之，则可以判断使用中继代理。

正如前面提到的租约的例子，租约是有一定时间的，对于 IP 地址分配也是一样，如图 2-56 所示，在协议中可以看到提供的地址是 10.135.232.233，它允许租用的时间是 7200s。

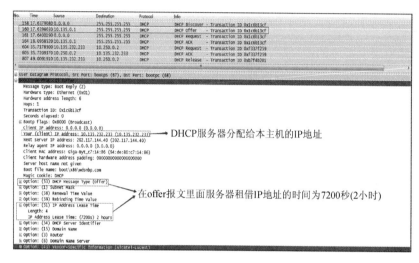

图 2-56 Wireshark 显示 DHCP 租约时间分析

以租约类比，如果租约到期证明不可以再继续使用这个地址，在网络中对于动态分配的地址，租约到期要进行释放。可以很容易想到，如果不释放，那么长此下去资源肯定是会造成浪费而且必然是不够使用的；如果释放出现意外，IP 地址则不能很好地同步。

当然，对于续约也是一样，DHCP Request 的数据包，目的是请求更新自己的租约。如果 DHCP 服务器正常且响应了此请求，那么就会返回一个 DHCP ACK 的数据包，这表示续约成功。如果第一次没有续约成功，到了租期的 7/8 时，还会重复一次申请续约的过程。如果成功，新的租期自然是在申请日期的基础上加上租约时间，依此类推。

在本章中学习了被 Internet 普遍采用的 C/S 模式，并主要介绍了处于应用层的 3 个协议，分别是 HTTP、DNS 和 DHCP，协议是网络中心的核心概念，对应用层协议的学习使对"协议是什么"有了更为直观的认识。这也将为之后学习其他章节打下基础。

第 3 章 运输层典型协议分析

3.1 运输层概述

OSI 七层模型中的物理层、数据链路层和网络层，它们是面向网络通信的低三层协议。运输层负责端到端的通信，既是七层模型中负责数据通信的最高层，又是面向网络通信的低三层和面向信息处理的最高三层之间的中间层。运输层位于网络层之上、会话层之下，它利用网络层子系统提供给它的服务去开发本层的功能，并实现本层对会话层的服务。

运输层向它上面的应用层（依据 TCP/IP 参考模型）提供运输服务，通信的真正端点从运输层的角度来看，并不是主机而是主机中的进程，因此也可以说运输层提供了进程到进程的逻辑通信服务。运输层的功能主要有多路复用（multiplexing）、多路分解（dumultiplexing）。本章重点分析的是运输层最重要的两种协议：用户数据报协议（user datagram protocl，UDP）和传输控制协议（transmission control protocol，TCP）。

3.2 TCP

3.2.1 TCP 简介

TCP 是一种面向连接的、可靠的、基于字节流的传输层通信协议，由 IETF 的 RFC 793 定义。在简化的计算机网络 OSI 模型中，它完成第四层运输层所指定的功能。在 Internet 协议族（TCP/IP）中，TCP 层是位于 IP 层之上、应用层之下的中间层。不同主机的应用层之间经常需要可靠的、像管道一样的连接，但是 IP 层不提供这样的流机制，而是提供不可靠的包交换，因此 TCP 在运输层提供可靠传输服务来保证数据的正确性。

由 TCP 传递给 IP 的信息单位称为报文段。当 TCP 发出一个报文段后，它启动一个定时器，等待目的端确认收到这个报文段。如果不能及时收到一个确认，它就重发这个报文段。当 TCP 收到发自 TCP 连接另一端的数据，它将发送一个确认。这个确认不是立即发送，通常延迟几分之一秒。TCP 首部具有校验和，目的是检测数据在传输过程中的任何变化，如果收到报文段的校验和有差错，TCP 将不确认收到并丢弃这个报文段。既然 TCP 报文段作为 IP 数据报来传输，因为 IP 数据报的到达可能失序，因此 TCP 报文段的到达也可能失序。如果必要，TCP 将对收到的数据进行排序，将收到的数据以正确的顺序交给应用层。而且 IP 数据报会发生重复，TCP 提供可靠性，连接端必须丢弃重复的数据。TCP 还能提供流量控制，TCP 连接的每一方都有固定大小的缓冲空间。TCP 的接收端只允许另一端发送接收端缓冲区所能接纳的数据。这将防止较快主机致使较慢主机的缓冲区溢出。

3.2.2　TCP 报文格式

TCP 数据被封装在一个 IP 数据报中，格式如图 3-1 所示。

16 位源端口号							16 位目的端口号
32 位序号							
32 位确认序号							
4位首部长度	保留6位	URG	ACK	PSH	RST	SYN	FIN	16 位窗口大小
16 位校验和							16 位紧急指针
选项							
数据							

图 3-1　TCP 报文结构

（1）每个 TCP 段都包括源端和目的端的端口号，用于寻找发送端和接收端的应用进程。这两个值加上 IP 首部的源端 IP 地址和目的端 IP 地址唯一确定一个 TCP 连接。

（2）序号用来标识从 TCP 发送端向接收端发送的数据字节流，它表示在这个报文段中的第一个数据字节。如果将字节流看成在两个应用程序间的单向流动，则 TCP 用序号对每个字节进行计数。

（3）当建立一个新连接时，SYN 标志变为 1。序号字段包含由这个主机选择的该连接的初始序号 ISN，该主机要发送数据的第一个字节的序号为这个 ISN 加 1，因为 SYN 标志使用了一个序号。

（4）既然每个被传输的字节都被计数，确认序号包含发送确认的一端所期望收到的下一个序号。因此，确认序号应当是上次已成功收到的数据字节序号加 1。只有 ACK 标志为 1 时确认序号字段才有效。

（5）发送 ACK 无需任何代价，因为 32 位的确认序号字段和 ACK 标志一样，总是 TCP 首部的一部分。因此一旦一个连接建立起来，这个字段总是被设置，ACK 标志也总是被设置为 1。

（6）TCP 为应用层提供全双工的服务。因此，连接的每一端必须保持每个方向上的传输数据序号。

（7）TCP 可以表述为一个没有选择确认或否认的滑动窗口协议。因此 TCP 首部中的确认序号表示发送方已成功收到字节，但还不包含确认序号所指的字节。当前还无法对数据流中选定的部分进行确认。

（8）首部长度需要设置，因为选项字段的长度是可变的。TCP 首部最多 60 字节。

（9）6 个标志位中的多个可同时设置为 1。
① URG——紧急指针有效；
② ACK——确认序号有效；
③ PSH——接收方应尽快将这个报文段交给应用层；
④ RST——重建连接；
⑤ SYN——同步序号用来发起一个连接；
⑥ FIN——发送端完成发送任务。

（10）TCP 的流量控制由连接的每一端通过声明的窗口大小来提供。窗口大小为字节数，起始于确认序号字段指明的值，这个值是接收端期望接收的字节数。窗口大小是一个 16 位的字段，因而窗口大小最大为 65535 字节。

（11）校验和覆盖整个 TCP 报文段。这是一个强制性的字段，一定由发送端计算和存储，并由接收端进行验证。TCP 校验和的计算和 UDP 首部校验和的计算方法一样。

（12）紧急指针是一个正的偏移量，序号字段中的值相加表示紧急数据最后一个字节的序号。TCP 的紧急方式是发送端向另一端发送紧急数据的一种方式。

（13）最常见的可选字段是最大报文段长度（MMS），每个连接方通常都在通信的第一个报文段中指明这个选项。它指明本端所能接收的最大长度的报文段。

3.2.3 TCP 连接建立与拆除

1. 建立连接

（1）请求端发送一个 SYN 段指明客户打算连接的服务器的端口，以及初始序号（ISN）。

（2）服务器端发回包含服务器的初始序号的 SYN 报文段作为应答。同时将确认序号设置为客户的 ISN 加 1 以对客户的 SYN 报文段进行确认。一个 SYN 将占用一个序号。

（3）客户必须将确认序号设置为服务器的 ISN 加 1 以对服务器的 SYN 报文段进行确认。这 3 个报文段完成连接的建立，称为三次握手。发送第一个 SYN 的一端将执行主动打开，接收这个 SYN 并发回下一个 SYN 的另一端执行被动打开。

2. 拆除连接

由于 TCP 连接是全双工的,所以每个方向都必须单独进行关闭。原则是当一方完成它的数据发送任务后就能发送一个 FIN 来终止这个方向的连接。收到一个 FIN 只意味着这一方向上没有数据流动,一个 TCP 连接在收到一个 FIN 后仍能发送数据。首先进行关闭的一方将执行主动关闭,而另一方执行被动关闭。

(1) TCP 客户端发送一个 FIN,用来关闭客户到服务器的数据传送。
(2) 服务器收到这个 FIN,发回一个 ACK,确认序号为收到的序号加 1。和 SYN 一样,一个 FIN 将占用一个序号。
(3) 服务器关闭客户端的连接,发送一个 FIN 给客户端。
(4) 客户段发回确认,并将确认序号设置为收到序号加 1。

3. TCP 的半关闭

TCP 提供了连接的一端在结束它的发送后还能接收来自另一端数据的能力,这就是 TCP 的半关闭。客户端发送 FIN,另一端发送对这个 FIN 的 ACK 报文段。当收到半关闭的一端在完成它的数据传送后,才发送 FIN 关闭这个方向的连接,客户端再对这个 FIN 确认,这个连接才彻底关闭。

4. 同时打开与关闭

为了处理同时打开,对于同时打开它仅建立一条连接而不是两条连接。两端几乎在同时发送 SYN,并进入 SYN_SENT 状态。当每一端收到 SYN 时,状态变为 SYN_RCVD,同时它们都再发 SYN 并对收到的 SYN 进行确认。当双方都收到 SYN 及相应的 ACK 时,状态都变为 ESTABLISHED。一个同时打开的连接需要交换 4 个报文段,比正常的三次握手多了一次。

当应用层发出关闭命令,两端均从 ESTABLISHED 变为 FIN_WAIT_1。这将导致双方各发送一个 FIN,两个 FIN 经过网络传送后分别到达另一端。收到 FIN 后,状态由 FIN_WAIT_1 变为 CLOSING,并发送最后的 ACK。当收到最后的 ACK,状态变为 TIME_WAIT。同时关闭和正常关闭的段减缓数目相同。

3.2.4 最大报文段长度

最大报文段长度(maximum segment size,MSS)表示 TCP 传往另一端的最大块数据的长度。当一个连接建立时,连接的双方都要通告各自的 MSS。

一般,如果没有分段发生,MSS 越大越好。报文段越大允许每个报文段传送的数据越多,相对 IP 和 TCP 首部有更高的网络利用率。当 TCP 发送一个

SYN 时，它能将 MSS 值设置为外出接口的 MTU 长度减去 IP 首部和 TCP 首部长度。对于以太网，MSS 值可达 1460。

如果目的地址为非本地的，MSS 值通常默认为 536，是否本地主要通过网络号区分。MSS 让主机限制另一端发送数据报的长度，加上主机也能控制它发送数据报的长度，这将使以较小 MTU 连接到一个网络上的主机避免分段。

3.2.5 TCP 的三次握手与四次挥手

1. 三次握手建立连接

（1）请求端（通常称为客户）发送一个 SYN 段指明客户打算连接的服务器的端口，以及初始序号（ISN，在这个例子中为 1415531521，如图 3-2 所示）。这个 SYN 段为报文段 1。

（2）服务器发回包含服务器的初始序号的 SYN 报文段（报文段 2）作为应答。同时将确认序号设置为客户的 ISN 加 1 以对客户的 SYN 报文段进行确认。一个 SYN 将占用一个序号。

（3）客户必须将确认序号设置为服务器的 ISN 加 1 以对服务器的 SYN 报文段进行确认。

这三个报文段完成连接的建立，这个过程又称三次握手（three-way handshake），如图 3-2 所示。

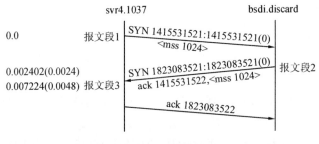

图 3-2 三次握手

用 Wireshark 三次握手抓包如图 3-3 所示。

```
192.168.0.18      192.168.0.145     TCP  54339 > ddi-tcp-1 [SYN] Seq=0 Win=5840 Len=0 MSS=1460 TS
192.168.0.145     192.168.0.18      TCP  ddi-tcp-1 > 54339 [SYN, ACK] Seq=0 Ack=1 Win=5792 Len=0
192.168.0.18      192.168.0.145     TCP  54339 > ddi-tcp-1 [ACK] Seq=1 Ack=1 Win=5856 Len=0 TSV=
```

图 3-3 用 Wireshark 三次握手抓包

可以看到三次握手确定了双方之间分组的序号、最大接收数据的大小（window）以及 MSS。MSS = MTU－IP 头－TCP 头，MTU 表示最大传输单元，在 IP 报文一般为 1500 字节（假定底层为以太网）。IP 首部和 TCP 首部在不带可选选项时都是 20 字节。这样，MSS＝1500－20－20 ＝ 1460。MSS 限制了 TCP

包携带数据的大小为 1460 字节，它表示当应用层向传输层提交数据通过 TCP 协议进行传输时，如果应用层的数据大于 MSS 就必须分段，分成多个段，逐个地发过去。例如，应用层一次性向传输层提交 4096 字节数据，这个时候通过 Wireshark 抓包结果如图 3-4 所示。

```
192.168.63.132    192.168.63.131    TCP    54094 > ddi-tcp-1 [SYN] Seq=0 Win=5840 Len=0
192.168.63.131    192.168.63.132    TCP    ddi-tcp-1 > 54094 [SYN, ACK] Seq=0 Ack=1 Win=
192.168.63.132    192.168.63.131    TCP    54094 > ddi-tcp-1 [ACK] Seq=1 Ack=1 Win=5856
192.168.63.132    192.168.63.131    TCP    54094 > ddi-tcp-1 [ACK] Seq=1 Ack=1 Win=5856
192.168.63.132    192.168.63.131    TCP    54094 > ddi-tcp-1 [ACK] Seq=1449 Ack=1 Win=58
192.168.63.132    192.168.63.131    TCP    54094 > ddi-tcp-1 [PSH, ACK] Seq=2897 Ack=1 W
```

图 3-4　Wireshark 显示传输报文

前三次是三次握手的过程，后面三次是传送数据的过程，由于数据大小是 4096 字节，所以用了三次进行传递（1448＋1448＋1200）。

在这里的 TCP 携带可选项，TCP 头长度 = 20 + 12（可选选项大小）= 32 字节。这样能传输的最大数据为 1500－20－32＝1448 字节。

2．四次挥手关闭连接

由于 TCP 连接是全双工的，所以每个方向都必须单独进行关闭。原则是当一方完成它的数据发送任务后就能发送一个 FIN 来终止这个方向的连接。收到一个 FIN 只意味着这一方向上没有数据流动，一个 TCP 连接在收到一个 FIN 后仍能发送数据。首先进行关闭的一方将执行主动关闭，而另一方执行被动关闭，如图 3-5 所示。

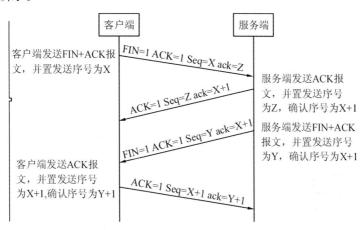

图 3-5　四次握手

Wireshar 显示 TCP 报文如图 3-6 所示。

```
192.168.0.145    192.168.0.18     TCP    ddi-tcp-1 > 54339 [FIN, ACK] Seq=1 Ack=1 Win=5792 Len=0 TSV=16
192.168.0.18     192.168.0.145    TCP    54339 > ddi-tcp-1 [ACK] Seq=1 Ack=2 Win=5856 Len=0 TSV=2380955
192.168.0.18     192.168.0.145    TCP    ddi-tcp-1 > 54339 [FIN, ACK] Seq=1 Ack=2 Win=5856 Len=0 TSV=23
192.168.0.145    192.168.0.18     TCP    ddi-tcp-1 > 54339 [ACK] Seq=2 Ack=2 Win=5792 Len=0 TSV=1634599
```

图 3-6　Wireshark 显示 TCP 报文

(1) TCP 客户端发送一个 FIN，用来关闭客户到服务器的数据传送。

(2) 服务器收到这个 FIN，它发回一个 ACK，确认序号为收到的序号加 1。和 SYN 一样，一个 FIN 将占用一个序号。

(3) 服务器关闭客户端的连接，发送一个 FIN 给客户端。

(4) 客户段发回 ACK 报文确认，并将确认序号设置为收到的序号加 1。

3.2.6 TCP 状态机

TCP 协议的操作可以使用一个具有 11 种状态的有限状态机（finite state machine）来表示，图 3-7 描述了 TCP 的有限状态机，图中的圆角矩形表示状态，箭头表示状态之间的转换，各状态的描述如表 3-1 所示。图中用粗线表示客户端主动和被动的服务器端建立连接的正常过程：客户端的状态变迁用粗实线，服务器端的状态变迁用粗虚线。细线用于不常见的序列，如复位、同时打开、同时关闭等。图中的每条状态变换线上均标有"事件/动作"：事件是指用户执行了系统调用（CONNECT、LISTEN、SEND 或 CLOSE）、收到一个报文段（SYN、FIN、ACK 或 RST），或者是出现了超过两倍最大的分组生命期的情况，事件的发生将执行某种动作并引起状态的变迁；动作是指发送一个报文段（SYN、FIN 或 ACK）或什么也没有（用"－"表示）。

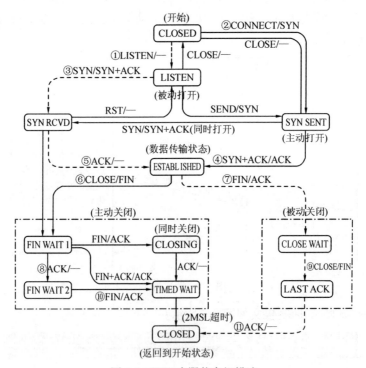

图 3-7 TCP 有限状态机描述

表 3-1 TCP 状态表

状态	描述
CLOSED	关闭状态，没有连接活动或正在进行
LISTEN	监听状态，服务器正在等待连接进入
SYN RCVD	收到一个连接请求，尚未确认
SYN SENT	已经发出连接请求，等待确认
ESTABLISHED	连接建立，正常数据传输状态
FIN WAIT 1	（主动关闭）已经发送关闭请求，等待确认
FIN WAIT 2	（主动关闭）收到对方关闭确认，等待对方关闭请求
TIMED WAIT	完成双向关闭，等待所有分组死掉
CLOSING	双方同时尝试关闭，等待对方确认
CLOSE WAIT	（被动关闭）收到对方关闭请求，已经确认
LAST ACK	（被动关闭）等待最后一个关闭确认，并等待所有分组死掉

如图 3-7 所示的 TCP 有限状态机：粗实线表示客户的正常路径；粗虚线表示服务器的正常路径；细线表示不常见的事件。

每个连接均开始于 CLOSED 状态。当一方执行了被动的连接原语 LISTEN 或主动的连接原语 CONNECT 时，它便会脱离 CLOSED 状态。如果此时另一方执行了相对应的连接原语，连接便建立起来，并且状态变为 ESTABLISHED。任何一方均可以首先请求释放连接，当连接释放后，状态又回到 CLOSED。

3.2.7 TCP 分析实验

在这个实验中，将对 TCP 进行详细研究。将研究 TCP 用于提供可靠数据传输的序号和确认号；我们将看到 TCP 拥塞控制算法，即慢启动和拥塞避免；还会简要考察 TCP 建立连接和之后的行为（吞吐量和往返时间）。

TCP 分析实验如下。

实验过程为：将包含着 Lewis Carrol 的小说 Alice's Adventures in Wonderland 的大小为 150KB 的文件，从实验者的计算机传输到远程服务器上，并对此过程进行数据抓包。具体抓包过程和分析如下。

（1）启动浏览器，进入网页 http：//gaia.cs.umass.edu/Wireshark-labs/alice.txt 并下载 "Alice's Adventures in Wonderland" 文件，存储在计算机上。

（2）访问 http：//gaia.cs.umass.edu/Wireshark-labs/TCP-Wireshark-file1.html，界面如图 3-8 所示。

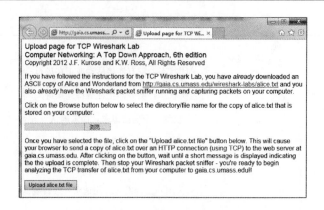

图 3-8　Wireshark 显示网站访问

（3）点击"浏览…"按钮，选择刚才下载存储的"Alice's Adventures in Wonderland"文件，但是不要按"Upload alice.txt file"按钮。

（4）启动 Wireshark 开始抓包（Capture→Start），然后在 Wireshark 抓包选项界面按 OK 键，此时在协议筛选框不填写任何选项。

（5）回到浏览器，按"Upload alice.txt file"按钮将上传文件到 gaia.cs.umass.edu server，一旦上传成功浏览器会显示成功信息。

（6）停止抓包，在协议筛选框输入"tcp"以便分析，本实验分析 TCP 而非 HTTP，所以通过选择 Analyze→Enabled Protocols 对 HTTP 取消筛选。

1. IP 地址和 TCP 端口号分析

如图 3-9 所示，第 199 个帧是客户端发送的 HTTP 用 POST 方法的请求报文段，源 IP（即客户机 IP）地址为 192.168.1.102，目的 IP（即服务器 IP）地址为 128.119.245.12；源端口号为 1161，目的端口号为 80。

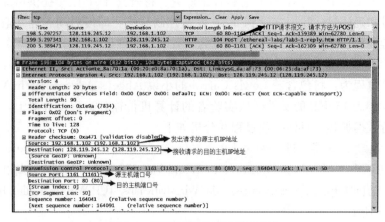

图 3-9　HTTP 的 POST 请求报文

如图 3-10 所示为三次握手的第一步，由客户机发送 SYN 报文段，其序号为 0，通过标志 SYN 位置 1 以确定。

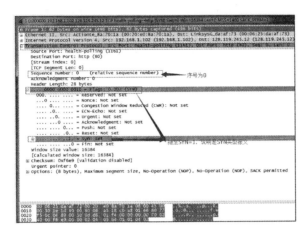

图 3-10　三次握手之 SYN

2. 三次握手分析

接下来分析第二步，客户机接收到 SYN/ACK 报文段，SYN/ACK 位置 1，序号为 0 确认号为 1，如图 3-11 所示。

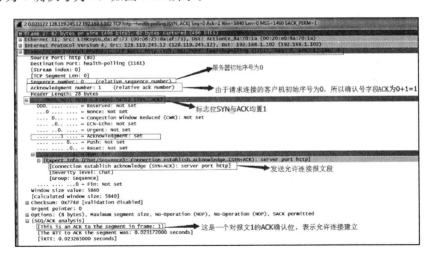

图 3-11　三次握手之 ACK

3. TCP 分段传输和拥塞控制分析

由图 3-12 可知，在文本传输过程中 TCP 进行了分段传输。

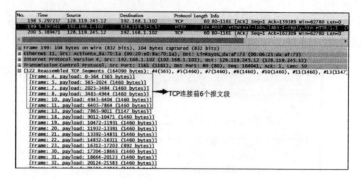

图 3-12 TCP 报文分段

如图 3-13 和图 3-14 所示，选取其前 6 个报文段，前 6 个 TCP 报文段长度分别为 565 字节，1460 字节，1460 字节，1460 字节，1460 字节，1460 字节。统计了每个报文段的发送时间和接收时间，统计方法如下。

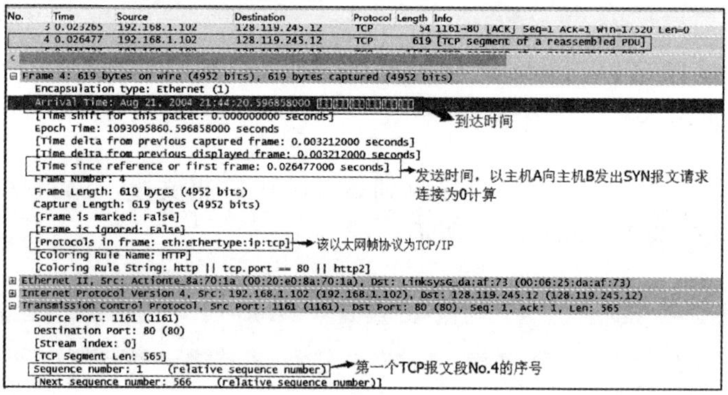

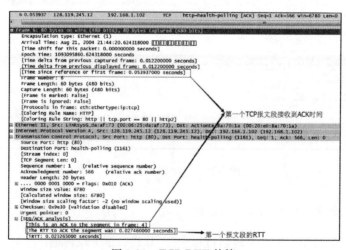

图 3-13 TCP 分段分析

图 3-14 TCP RTT 估算

计算 SampleRTT、EstimatedRTT 方法如下。

Estimated RTT ＝（1－α）· EstimatedRTT＋α · SampleRTT，通常 α 取 0.125。结果如表 3-2 所示。

表 3-2 计算 SampleRTT 和 EstimatedRTT

Tcp segment	Send time/s	Receive time/s	SampleRTT/s	EstimatedRTT/s
Segment1	0.026477	0.053937	0.027460	0.027460
Segment2	0.041737	0.077294	0.035557	0.028472
Segment3	0.054026	0.124085	0.070059	0.033670
Segment4	0.054690	0.169118	0.114428	0.043765
Segment5	0.077405	0.217299	0.139894	0.055781
Segment6	0.078157	0.267802	0.189645	0.072514

接下来计算 TCP 的平均吞吐量，选择整个连接时间作为平均时间段。然后，TCP 连接的平均吞吐量为总的传输数据与总传输时间的比值。传输数据总量为序号为 TCP 报文段第一个序列号（第 3 段 1B）和最后一个序列号的 ACK（第 202 段 164091B）的差值，总数据量为 164091－1＝164090B。总时间为第一个 TCP 段的时间（第 4 段时间为 0.026477s）和最后一个 ACK 的时间（第 202 段时间为 5.45583s）的差值，所以总时间为 5.455830－0.026477＝5.429353s，平均吞吐量 average throughput＝164090/5.429353/1024＝29.5144 KB/s。详见图 3-15 和图 3-16。

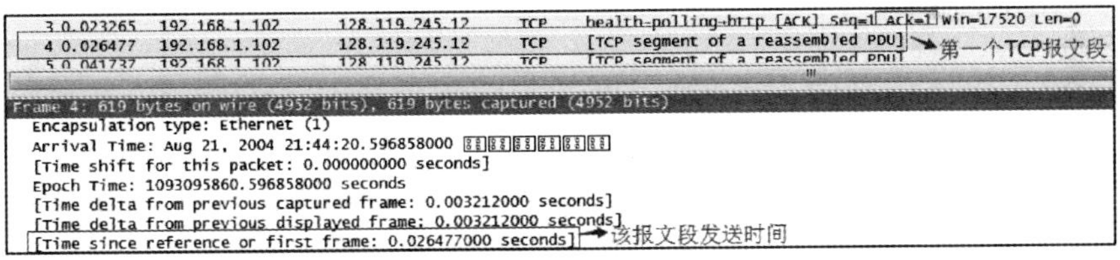

图 3-15 Wireshark 显示报文段发送时间信息

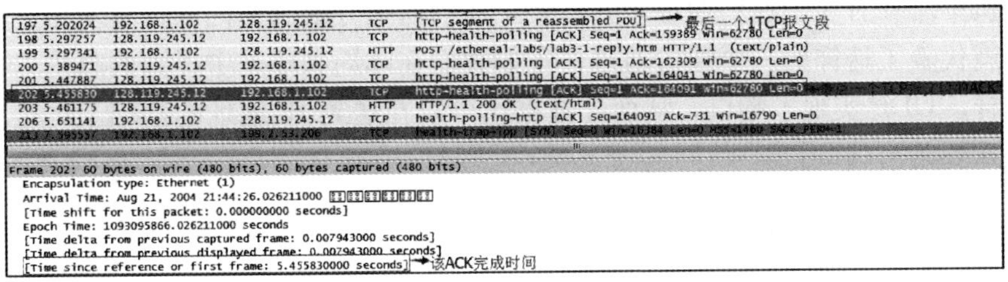

图 3-16 Wireshark 显示报文段 ACK 时间信息

最后在菜单栏选择 Statistics→TCP Stream Graph→Time-Sequence-Graph

(Stevens)，可以得到如图 3-17 所示结果。在时序（Stevens）图中，可以看到序号先是指数增长，然后是线性增加，分别对应 TCP 拥塞控制的慢启动阶段和拥塞避免阶段。

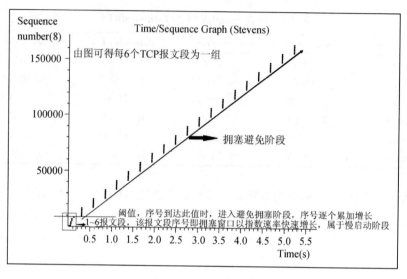

图 3-17　TCP 的拥塞控制

3.3　UDP

3.3.1　UDP 简介

UDP 传输与 IP 传输非常类似。可以将 UDP 看成 IP 协议暴露在传输层的一个接口。UDP 同样以数据包（datagram）的方式传输，它的传输方式也是"best effort"的，所以 UDP 也是不可靠的。那么，为什么不直接使用 IP 协议而要额外增加一个 UDP 呢？一个重要的原因是 IP 中并没有端口（port）的概念。IP 协议进行的是 IP 地址到 IP 地址的传输，这意味着两台计算机之间的对话。但每台计算机中需要有多个通信通道，并将多个通信通道分配给不同的进程使用。一个端口就代表了这样的一个通信通道，正如我们在邮局和邮差中提到的收信人的概念一样。UDP 实现了端口，从而让数据包可以在送到 IP 地址的基础上，进一步可以送到某个端口。

3.3.2　UDP 基本工作原理

（1）UDP 用户数据报传输过程中的封装与拆封如图 3-18 所示。

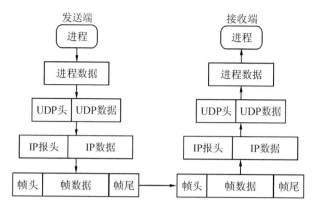

图 3-18　UDP 的封装与拆封

（2）UDP 报文传输队列如图 3-19 所示。

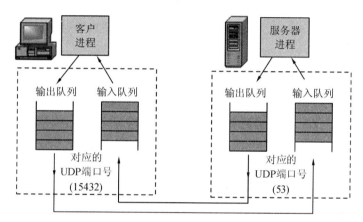

图 3-19　UDP 报文传输队列

（3）UDP 的复用和解复用如图 3-20 所示。

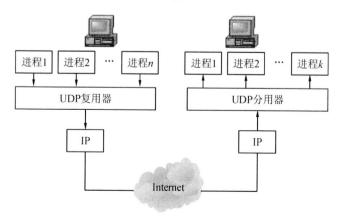

图 3-20　UDP 的复用和解复用

(4) TCP/IP 协议族中用端口号来标识进程。

端口号是在 0~65535 之间的整数，客户程序随机选取临时端口号，每一种服务器程序被分配了确定的全局一致的周知端口号，每一个客户进程都知道相应的服务器进程的周知端口号。

UDP 使用的周知端口号如表 3-3 所示。

表 3-3 UDP 的周知端口号

端口号	服务进程	说明
53	Name sever	域名服务
67	Bootps	下载引导程序信息的服务器端口
68	Bootpc	下载引导程序信息的客户机端口
69	TFTP	简单文件传输协议
111	RPC	远程过程调用
123	NTP	网络时间协议
161	SNMP	简单网络管理协议

3.3.3 UDP 报文格式

UDP 报文分为首部字段和数据字段，其中首部字段只占用 8 字节，分别是各占用 2 字节的源端口、目的端口、长度及校验和，如图 3-21 所示。

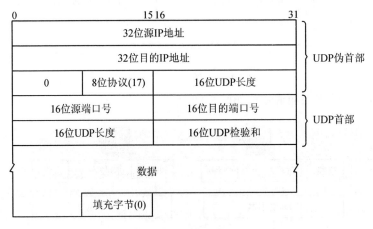

图 3-21 UDP 首部

长度：UDP 报文的整个大小，最小为 8 字节（仅为首部）。

校验和：在进行校验和计算时，会添加一个伪首部一起进行运算。

伪首部（占用 12 字节）：4 字节的源 IP 地址、4 字节的目的 IP 地址、1 字节的 0、1 字节的数字 17，以及占用 2 字节 UDP 长度。这个伪首部不是报文的真正首部,只是引入为了计算校验和。相对于 IP 协议的只计算首部,UDP 校验和会把首部和数据一起进行校验。接收端进行的校验和与 UDP 报文中的校验和相与,如果无差错应该全为 1。如果有误,则将报文丢弃或者发给应用层、并附上差错警告。

3.3.4 UDP 分析实验

UDP 实验过程不唯一,任意采用一种方法,只要实验者的计算机向服务器发送 UDP 数据包即可,本分析采用的是 DNS 查询方法。

具体过程如下。

在 DOS 窗口运行 "nslookup www.baidu.com 114.114.114.144" 进行 DNS 查询的行为,来对运输层 UDP 报文进行分析并运用 Wireshark 抓包分析。

(1) 任选一个 UDP 数据报,如图 3-22 所示,UDP 首部一共有四个字段,分别为源端口号、目的端口号、长度以及校验和,它们各占 2 字节,所以 UDP 首部一共占 8 字节。

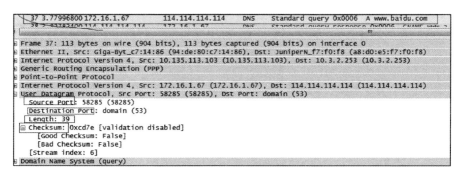

图 3-22 传输 DNS 的 UDP 报文段

(2) UDP 的最大端口号以及数据传输值。UDP 最大可能端口号理论值为 65535,即最大理论能传输的数据大小为 65536－1＝65535,除去 8 字节的首部,得到 65535－8＝65527。显然,只考虑运输层不考虑网络层是不对的。以太网数据(即 IP 数据报)长度必须在 46～1500 字节之间,这是由它的物理特性决定的。UDP 报文属于运输层,传输时,网络层(IP)将运输层报文(即 UDP 报文)封装到 IP 数据报中,所以 UDP 最大传载数据由 IP 数据报文和 UDP 报文确定,UDP 最大传载＝1500(以太网)－20(IP 数据报首部)－8(UDP)＝1472。如果再考虑 Internet 上的标准 MTU 值为 576 字节,建议在进行 Internet 的 UDP 编程时,最好将 UDP 的数据长度控件在 548 字节(576-8-20)以内。

(3) UDP 在 IP 数据报中的协议号为 17,如图 3-23 所示。

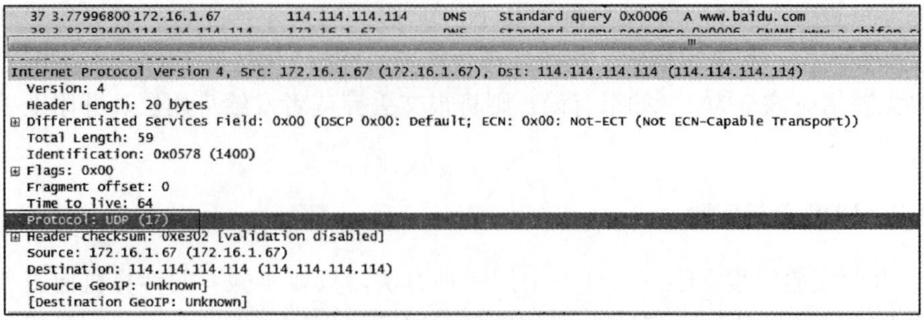

```
37 3.77996800 172.16.1.67        114.114.114.114    DNS    Standard query 0x0006  A www.baidu.com
```

```
Internet Protocol Version 4, Src: 172.16.1.67 (172.16.1.67), Dst: 114.114.114.114 (114.114.114.114)
    Version: 4
    Header Length: 20 bytes
  ⊞ Differentiated Services Field: 0x00 (DSCP 0x00: Default; ECN: 0x00: Not-ECT (Not ECN-Capable Transport))
    Total Length: 59
    Identification: 0x0578 (1400)
  ⊞ Flags: 0x00
    Fragment offset: 0
    Time to live: 64
    Protocol: UDP (17)
  ⊞ Header checksum: 0xe302 [validation disabled]
    Source: 172.16.1.67 (172.16.1.67)
    Destination: 114.114.114.114 (114.114.114.114)
    [Source GeoIP: Unknown]
    [Destination GeoIP: Unknown]
```

图 3-23　UDP 在 IP 数据报中的协议号为 17

第 4 章 网络层典型协议分析

网络层是构建互联网的基础，分组转发和动态路由选择是网络层的重要功能。Internet 网络层提供无连接、不可靠但尽力而为的分组传送服务，实现这种服务的主要协议是 IP 协议，ICMP、NAT 等用户辅助功能，本章将围绕这三个协议进行学习。

4.1 网络层简介

在进行网络层协议分析之前，有必要对网络层中一些基本概念进行简单的介绍。

如图 4-1 给出了一个具有 H_1 和 H_2 两台主机，且在 H_1 与 H_2 之间的路径上有几台路由器的简单网络。假设 H_1 正在向 H_2 发送信息，考虑这些主机与中间路由器中的网络层起的作用，H_1 中的网络层接收来自 H_1 运输层的每个报文段，将其封装成一个数据报，然后将数据报向相邻路由器 R_1 发送。在接收方主机 H_2，其网络层接收来自相邻路由器 R_2 的数据报，提取出运输层报文段，将其向上交付给 H_2 的运输层。路由器的主要作用便是将数据报从入数据链路层发到出链路。注意到图 4-1 中所提示的路由器具有删减的协议栈，即没有网络层以上的部分，因为路由器不运行在应用层和运输层。

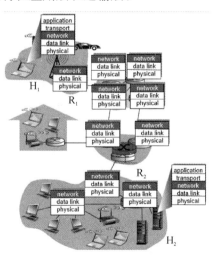

图 4-1 网络层结构示意图

4.1.1 转发和选路

简单来说网络层的作用，就是将分组从一台发送主机移动到另一台接收主机。为此，需要两种重要的网络层功能：转发和选路。

(1) 转发。当一个分组到达某路由器的一条输入链路时，该路由器必须将该分组移动到适当的输出链路。例如，图 4-1 中，来自主机 H_1 的路由器 R_1 的一个分组，必须向 H_2 路径上的下一台路由器转发。转发是指将分组从一个输入链路接口转移到适当的输出链路接口的路由器本地动作。

(2) 选路。当分组从发送方流向接收方时，网络层必须决定这些分组所采用的路由或路径。计算这些路径的算法称为选路算法，例如，一个选路算法将决定分组从 H_1 到 H_2 所遵循的路径。选路是指分组从源到目的地时，决定端到端路径的网络范围的进程。

如图 4-2 所示，描述了一个首部字段值为 0111 的分组到达路由器的过程。该路由器在它的转发表中索引，决定该分组的输出链路接口是接口 2，该路由器则在内部将该分组转发到接口 2。

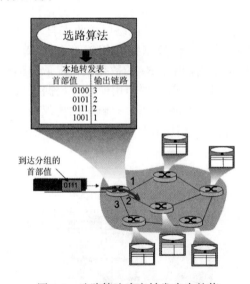

图 4-2 选路算法决定转发表中的值

4.1.2 分组交换

分组跨越互联网的传输基于分组交换技术，它广泛用于交换式网络，如互联网和广域网。分组交换是一种数据交换技术，规定如何将数据由源站经由交换节点转发到目的节点的方式，主要有两种基本方式。

(1) 数据报方式。每个路由器都独立寻径，它们可能经过不同路径和不同的传输时间，因此不能保证分组按顺序到达目的站。IP 数据包的转发就是基于这种方式。数据报网络如图 4-3 所示。

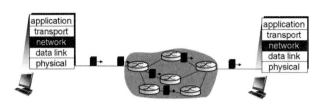

图 4-3　数据报网络

（2）虚电路方式。一次通信中所有分组都使用同一路径传输，为此，首先要建立一个传输连接，而后使用这一固定的路径进行数据传输，传输完成后释放这一连接，这与电路交换相似。但又有不同之处，虚电路连接并不是真实地建立一条物理线路，而是在现有的网络中指定一条传输路径，因此称为虚电路，而且这一条电路不是专用的，连接上的节点和线路还可以为转发其他传输的分组服务。虚电路网络如图 4-4 所示。

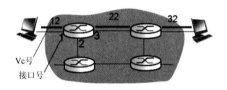

图 4-4　虚电路网络

4.1.3　路由器

实现网络互连的关键设备是路由器，路由器工作在网络层，其主要功能有两个：一是根据目的地址和路由表在它所连接的子网间进行分组转发；二是进行路由选择、动态优化和更新路由表。

图 4-5 显示了一个通用路由器体系结构的总体视图。其中表示了一台路由器的 4 个组成部分：输入端口、输出端口、交换结构、选路处理器。

（1）输入端口。输入端口要执行将一条输入的物理链路端接到路由器的物理层功能，它也要执行需要与位于如数据链路层远端的数据链路层功能交互的数据链路层功能，它还要完成查找与转发功能，以便转发到路由器交换结构部分的分组能出现在适当的输出端口。控制分组从输入端口到选路处理器。

（2）交换结构。交换结构将路由器的输入端口连接到它的输出端口，交换结构完全包含在路由器中，即它是一台网络路由器中的网络。

（3）输出端口。输出端口存储经过交换结构转发给它的分组，并将这些分组传输到输出链路。因此输出端口执行与输入端口顺序相反的数据链路层和物理层功能，当一条链路是双向链路时，与链路相连的输出端口通常与输入端口在同一线路卡上成对出现。

（4）选路处理器。选路处理器执行选路协议，维护选路信息与转发表，并执行路由器中的网络管理功能。

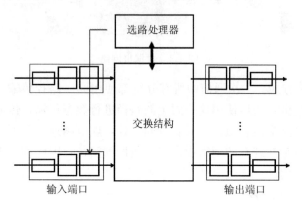

图 4-5　路由器体系结构

同时，路由器技术也在不断发展，功能更强的网络接口可以包含一个 CPU，具备网络层查找路由表和分组转发的功能，也使得路由器能更好更快地工作。

4.2　网际协议 IPv4

在本节中，集中学习在 Internet 是如何通过 IP 协议完成存储和转发的，目前有两个版本的 IP 协议在被使用，即 IPv4 和 IPv6，在这里重点研究广泛使用的 IP 协议版本 4，目前 IP 版本 6 已经被提议替代 IPv4。

4.2.1　IP 地址的分类及划分子网

IP 是英文 internet protocol 的缩写，是"网络之间互连的协议"，在 Internet 中，它规定了计算机在 Internet 上进行通信时应当遵守的规则。任何厂家生产的计算机系统，只要遵守 IP 协议就可以与 Internet 互联互通。因此，IP 协议也可以称为"因特网协议"。

IP 地址被用来给 Internet 上的计算机编址。每台联网的个人计算机上都需要有 IP 地址，才能正常通信。如果把"个人计算机"比作"一台电话"，那么"IP 地址"就相当于"电话号码"，而 Internet 中的路由器，就相当于电信局的"程控式交换机"。

IP 地址是一个 32 位的二进制数，通常被分割为 4 个"8 位二进制数"（4 字节)。IP 地址通常用"点分十进制"表示成(a.b.c.d)的形式,其中,a、b、c、d 都是 0~255 之间的十进制整数。例如,点分十进 IP 地址(100.4.5.6),实际上是 32 位二进制数。

IP 地址编址方案将 IP 地址空间划分为 A、B、C、D、E 五类,其中 A、B、C 是基本类,D、E 类作为多播和保留使用。基本 IP 地址结构如图 4-6 所示。

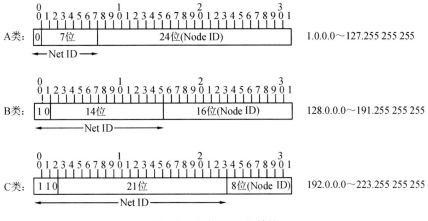

图 4-6 基本 IP 地址结构

IP 地址的范围如表 4-1 所示。

表 4-1 IP 地址范围

类别	最大网络数	网络号范围	最大主机数	主机号范围	IP 地址范围
A	126	1~126	16777214	0.0.1~255.254.254	1.0.0.1~126.255.254
B	16328	128.1~191.254	65534	0.1~255.254	128.1.0.1~191.254.255.254
C	2097150	192.0.2~223.225.254	254	1~254	192.0.0.1~223.255.254.254

IPv4 有 4 段数字,每一段最大不超过 255。由于互联网的蓬勃发展,IP 位址的需求量越来越大,IP 位址的发放越趋严格,有资料显示全球 IPv4 位址已经在 2011 年 4 月全部发完。

地址空间的不足必将妨碍互联网的进一步发展。为了扩大地址空间,拟通过 IPv6 重新定义地址空间。IPv6 采用 128 位地址长度。在 IPv6 的设计过程中除了一劳永逸地解决了地址短缺问题,还考虑了在 IPv4 中解决不好的其他问题。

4.2.2 子网掩码

子网掩码(subnet mask)又称网络掩码、地址掩码、子网络遮罩,它是一种用来指明一个 IP 地址的哪些位标识的是主机所在的子网,以及哪些位标识的是主机的位掩码。子网掩码不能单独存在,它必须结合 IP 地址一起使用。有几个常用的掩码，如 255.0.0.0：/8（A 类地址默认掩码）、掩码 255.255.0.0：/1（B 类地址默认掩码）、掩码 255.255.255.0：/24（C 类地址默认掩码）。子网掩码只有一个作用，就是将某个 IP 地划分成网络地址和主机地址两部分。图 4-7 给出了一个主机接入 Internet 时需要填写和配置的几个参数。

图 4-7 网络参数配置界面

4.2.3 IP 数据报格式

IP 数据报简称数据报，其格式如图 4-8 所示，分报头和数据区两部分，各字段解释如下。

32位		
ver / head. len / type of service		length
16bitidentifier	flgs	fragment offset
time to live	upper layer	header checksum
32bit source IP address		
32bit destination IP address		
options(if any)		
data (variable length, typically a TCP or UDP segment)		

图 4-8 IP 数据报格式

(1) 版本 (ver) 占 4 位，指 IP 协议的版本。通信双方使用的 IP 协议版本必须一致。目前广泛使用的 IP 协议版本号为 4（即 IPv4）。

(2) 首部长度 (head.len) 占 4 位，可表示的最大十进制数值是 15。请注意，这个字段所表示数的单位是 32 位字长（1 个 32 位字长是 4 字节），因此，当 IP 的首部长度为 1111 时（即十进制的 15），首部长度就达到 60 字节。最常用的首部长度就是 20 字节（即首部长度为 0101）。

(3) 区分服务 (type of service) 占 8 位，用来获得更好的服务。这个字段在旧标准中称为服务类型，但实际上一直没有被使用过。只有在使用区分服务时，这个字段才起作用。

(4) 总长度 (length) 指首部和数据之和的长度，单位为字节。总长度字段为 16 位，因此数据报的最大长度为 $2^{16}-1=65\,535$ 字节。

在 IP 层下面的每一种数据链路层都有自己的帧格式，其中包括帧格式中的数据字段的最大长度，称为最大传送单元（maximum transfer unit，MTU）。当一个数据报封装成数据链路层的帧时，此数据报的总长度（即首部加上数据部分）一定不能超过下面的数据链路层的 MTU 值。

(5) 标识 (identifier) 占 16 位。IP 软件在存储器中维持一个计数器，每产生一个数据报，计数器就加 1，并将此值赋给标识字段。但这个"标识"并不是序号，因为 IP 是无连接服务，数据报不存在按序接收的问题。当数据报由于长度超过网络的 MTU 而必须分片时，这个标识字段的值就被复制到所有的数据报的标识字段中。相同的标识字段的值使分片后的各数据报片最后能正确地重装成为原来的数据报。

(6) 标志 (flgs) 占 3 位，但目前只有 2 位有意义。

标志字段中的最低位记为 MF (more fragment)。MF=1 表示后面"还有分片"的数据报。MF=0 表示这已是若干数据报片中的最后一个。

标志字段中间的一位记为 DF (don't fragment)，意思是"不能分片"。只有当 DF=0 时才允许分片。

(7) 片偏移 (fragment offset) 占 13 位。片偏移指出：较长的分组在分片后，某片在原分组中的相对位置。也就是说，相对用户数据字段的起点，该片从何处开始。片偏移以 8 字节为偏移单位。也就是说，每个分片的长度一定是 8 字节（64 位）的整数倍。

(8) 生存时间 (time to live) 占 8 位，生存时间字段常用的英文缩写是 TTL (time to live)，表明是数据报在网络中的寿命。由发出数据报的源点设置这个字段。其目的是防止无法交付的数据报无限制地在 Internet 中存在，因而白白消耗网络资源。每经过一个路由器时，就把 TTL 减去数据报在路由器消耗掉的一段时间。当 TTL 值为 0 时，就丢弃这个数据报。

(9) 协议 (upper layer) 占 8 位，协议字段指出此数据报携带的数据是使用何种协议，以便使目的主机的 IP 层知道应将数据部分上交给哪个运输层协议。

(10) 首部校验和 (header checksum) 占 16 位。这个字段只校验数据报的首部，但不包括数据部分。这是因为数据报每经过一个路由器，路由器都要重新计算一下首部校验和（一些字段，如生存时间、标志、片偏移等都可能发生变化）。

(11) 源地址和目的地址占 32 位。

如图 4-9 展示了在 Wireshark 下抓包获得的 IP 报文格式分析。

图 4-9　Wireshark 下抓包 IP 报文格式

4.2.4　IP 数据报的分片、重组及转发

1. 分片

从传输层-网络层角度看，传输层数据加上 IP 数据报头的总长度必须小于 65 535 字节，如果大于，那么就需要把传输层数据分别封装在不同的 IP 数据报中。由于多个数据包传输，出错率增高，所以 TCP/IP 协议在应用层和传输层就已经开始控制报文的长度，以避免被分成多个数据报。

从网络层-数据链路层角度看，由于 IP 数据的最大长度为 65 535 字节，那么它所使用的网络的数据链路层最大传输单元的长度是 65 535 字节，那么传输效率一定很高，但是，实际上大量使用的网络的 MTU 都小于 IP 数据包的最大长度。因此，使用这些网络传输 IP 数据报时，要把 IP 数据包分成若干较小的片来传输，是否进行分片一般由路由器决定。

2. 重组

片重组是分片的逆过程，但片重组只在目的站进行，中途路由器不进行重组,这种重组方式简化了路由器的处理。但也有缺点:首先,如果数据报先通过 MTU 小的网络,而接下去又经过 MTU 大的网络,将导致网络宽带的浪费;其次,一个分片的丢失将导致整个数据报不能重组,因此,分片越多整个数据报丢失的概率越大。图 4-10 所示为数据报的分片和重组。

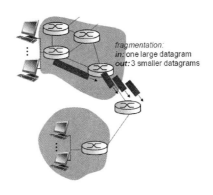

图 4-10　IP 数据包的分片与重组

如图 4-11 给出了 Wireshark 下 IP 分片的数据显示。

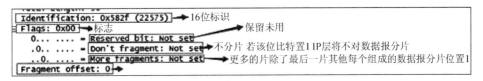

图 4-11　Wireshark 下 IP 分片显示

3. 转发

1) IP 数据包的直接交付和间接交付

直接交付是指源站或路由器将数据报直接传送到目的站,中间不需要其他路由器转发,否则就是间接交付。具体转发流程如下。

(1) IP 数据包到达网络层之后,首先根据目的 IP 地址得到目的网络号,然后决定是直接交付还是转发数据包。如果网络号不匹配,需要转发数据包,则跳到下一跳。

(2) 将数据包转发给目的主机。

(3) 首先根据目的 IP 地址在路由表(转发表)中查找下一跳 IP 地址。

(4) 然后在路由器的 ARP 高速缓存表中查找下一跳 IP 地址对应的 MAC 地址,如果找到下一跳路由器的 MAC 地址,则将查到的 MAC 地址填入数据帧的首部 6 字节(即更新数据链路层的数据帧);如果 ARP 高速缓存表中不存在此 IP 地址,则通过向当前局域网内广播一个 ARP 分组来请求下一跳路由器的 MAC 地址。ARP 请求分组广播出去后,只有下一跳路由器会对此请求分组做出响应，所有其他的主机和路由器都将忽略此 ARP 广播分组。

（5）根据得到的下一跳路由器 MAC 地址来更新数据链路层的数据帧，即帧头的目的 MAC 地址字段。

（6）转发数据包。

2）基本的 IP 数据报转发算法

（1）默认路由。IP 转发中常常使用默认路由。IP 首先在路由表中查找目的网络，如果表中没有相应的路由，则把数据报发给一个默认路由器。

（2）特定主机路由。路由表一般使用目的主机所在的网络而不是单个主机，但作为特例，IP 也允许使用完整的 IP 地址指定某个目的主机的路由，称为特定主机路由。

（3）统一的 IP 数据报转发算法。子网 IP 数据报转发机制，即为了将分组在子网中进一步转发，基本路由表需增加子网掩码

（目的网络 IP 地址，子网掩码，下一跳 IP 地址）

如果允许任意形式的子网掩码，子网 IP 数据报转发算法就可以兼容基本的 IP 数据报转发算法，得到统一形式的算法，为此，对子网掩码形式进行进一步规定如下。

① 划分子网的网络，子网掩码规定不变。

② 不划分子网的网络，其子网掩码形式规定为 IP 地址的 host-id 部分对应的位为 "0"，其余位为 "1"。

③ 特定主机路由，子网掩码规定为全 "1"。

④ 默认路由，其 IP 地址为 0.0.0.0，子网掩码为全 "0"。

如图 4-12 所示，给出了一个 IP 编址与接口的例子，在该图中，一台路由器用于互连七台主机，在它们的 IP 地址中，前 24 位相同，互连这三台主机的接口

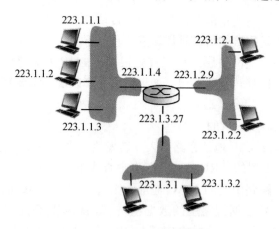

图 4-12 接口地址和子网地址

与路由器的一个接口的网络形成一个子网。IP 为这个子网分配一个地址 223.1.1.0/24，其中/24 记法有时称为子网掩码，它表明 32 位中的最左侧 24 位定义了子网地址。可以容易看出，图中给出了 3 个 IP 子网。

4.2.5 IP 协议分析实验

在这个实验中，将分析 IP 协议的各个方面，把重点放在 IP 数据报报文格式上。通过利用 traceroute 程序发送不同长度的数据报追踪分析 IP 数据包的接收和发送，来对 IP 协议进行分析。

在 Windows 系统中，Windows 提供的 tracert 不允许改变 ICMP 回送请求的大小，一个更好的 traceroute 工具是 pingplotter，可以通过 http://www.pingplotter.com 下载并安装 pingplotter。ICMP 回送请求报文的大小可以通过选择菜单项：Edit→Options→Packet Options 填写分组字段大小，显式地设置在 pingplotter 中，默认数据包大小为 56 字节。

1. 实验一：通过 traceroute 捕获数据

按如下步骤进行抓包。

（1）启动 Wireshark 开始包捕获（Capture→Start），在 Wireshark 抓包控件中选择"OK"（在这里不需要选择其他任何选项）。

（2）如果使用的是 Windows 平台，启动 pingplotter，在"Address to Trace Window"中输入一个目标地址，在"♯ of times to trace"中输入 3。选择 Edit→Advanced Options→Packet Options 并在 Packet Size Field 中输入值 56，然后按确定按钮进行抓包。可以看到一个 pingplotter 窗口如图 4-13 所示。

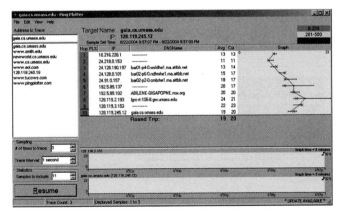

图 4-13　pingplotter 工具抓包显示

（3）接下来，发送一组长度较长的数据报，通过选择 Edit→Advanced Options→Packet Options 然后输入值为 2000，点击 OK，然后按"恢复"按钮。

（4）最后，发送一组长度更长的数据报，通过选择 Edit→Advanced Options→Packet Options 然后输入值为 3500，点击 OK，然后按"恢复"按钮。

（5）停止 Wireshar 抓包。

对实验获取的内容进行分析。在一系列 ICMP 回送请求消息中，选择第一个主机发送的 ICMP 回送请求消息，如图 4-14 所示，在 IP 协议中 src 处看到主机 IP 地址，在图中为 10.135.86.113，注意到 Header Length 和 Total Length 的值分别为 20 和 56，表示 IP 数据报的首部字段为 20 位，总长度为 56 位。需要说明的是，20 位表示 IP 首部，这个长度是不会改变的，但是数据报的总长度是可以改变的，如 100、200 等，接下来看到上层协议的 value 值为 1，表示上层协议是 ICMP 协议，并且注意到 Flag 字段的值为 0x00，表示该数据报未分片。而且 Fragment offset（偏移量）也为 0。

图 4-14　Wireshark 下第一个 IP 数据报报文

2. 实验二：IP 数据包的分片

接下来对 IP 数据报内容进行排序，单击 column header，将会进行排序。选择第一个由你的计算机发送的 ICMP 回送请求消息，打开协议的详细内容。在"列表数据包"窗口，你也将看到所有随后的 ICMP 消息。具体如图 4-15～图 4-18 所示。

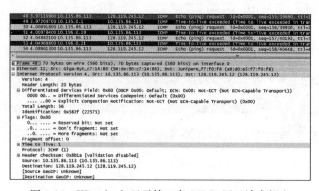

图 4-15　Wireshark 显示第一条 ICMP 回显请求报文

图 4-16 Wireshark 显示第二条 ICMP 回显请求报文

图 4-17 Wireshark 显示第三条 ICMP 回显请求报文

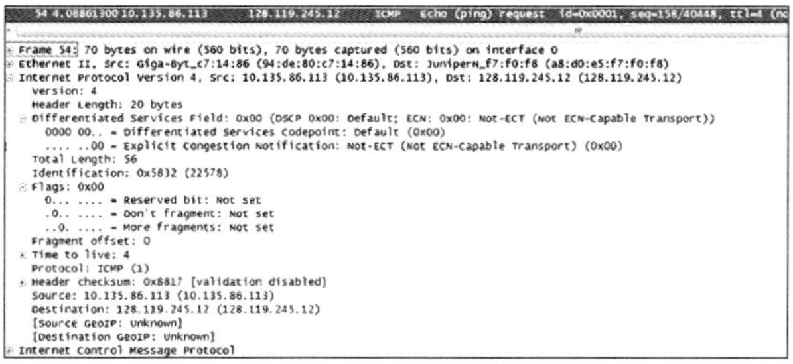

图 4-18 Wireshark 显示第四条 ICMP 回显请求报文

将上述 Wireshark 报文用表 4-2 将所需数据整理，见表 4-2。

表 4-2 未分片时数据报文信息

ID	Version	Header length	Type of service	Total length	Identification	Flags	Fragment offset	TTL	Header checksum	Source	Destination
48	4	20	0x00	56	0x582f	0x00	0	1	0x8b1a	10.135.86.113	128.119.245.12
50	4	20	0x00	56	0x5830	0x00	0	2	0x8a19	10.135.86.113	128.119.245.12
52	4	20	0x00	56	0x5831	0x00	0	3	0x8919	10.135.86.113	128.119.245.12
54	4	20	0x00	56	0x5832	0x00	0	4	0x8817	10.135.86.113	128.119.245.12
⋮	⋮	⋮	⋮	⋮	⋮	⋮	⋮	⋮	⋮	⋮	⋮

可以看出：Identification、TTL 和 Header checksum 字段在数据报中总是改变的，其中 Version、Header length、Type of service、Total length、Flags、Fragment offset、Source 和 Destination 字段保持不变，Version、Header length、Source 和 Destination 字段必须保持不变，Identification 和 checksum 字段必须改变。在数据报中 Identification 的值逐渐增大，每次增加十六进制的一个单位。

很明显表 4-2 显示的情况是在 56 字节发生的情况，表 4-3 显示了 2000 字节分析数据报分片情况。看到 Total length、Flags、Fragment offset、Header checksum"等字段"在两个报文中是变化的。

表 4-3　数据报长度为 2000 时的数据分析

ID	Version	Header length	Type of service	Total length	Identification	Flags	Fragment offset	TTL	Upperlayer protocol	Header checksum	Source	Destination
408	4	20	0x00	1500	0x58b0 (22704)	0x01	0	1	ICMP	0x64f5	10.135.86.113	128.119.245.12
409	4	20	0x00	520	0x58b0 (22704)	0x00	1480	1	ICMP	0x8810	10.135.86.113	128.119.245.12

当数据报长度为 3500 时的数据时，可以看出和数据报为 2000 相似，Total length、Flags、Fragment offset、Header checksum"等字段"在报文中是变化的，如表 4-4 所示。

表 4-4　数据报长度为 3500 时的数据分析

ID	Version	Header length	Type of service	Total length	Identification	Flags	Fragment offset	TTL	Upperlayer protocol	Header checksum	Source	Destination
604	4	20	0x00	1500	0x58d6 (22742)	0x01	0	1	ICMP	0x64cf	10.135.86.113	128.119.245.12
605	4	20	0x00	1500	0x58d6 (22742)	0x01	1480	1	ICMP	0x6416	10.135.86.113	128.119.245.12
606	4	20	0x00	540	0x58d6 (22742)	0x00	2960	1	ICMP	0x871d	10.135.86.113	128.119.245.12

接下来探讨一下离计算机最近的第一条路由器的 TTL 值的情况，如图 4-19 所示。注意到在数据报中 TTL 的值为 255，这表示第一跳路由器的值，这个 TTL 值是恒定的，之后每经过一跳，TTL 的值将减一。

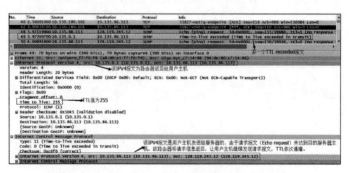

图 4-19　Wireshark 下第一跳路由器 TTL 分析

第 4 章　网络层典型协议分析　　　　　　　　　　　　　　　　• 73 •

在实验一中可以看到，当数据长度为 56 字节时偏移量标志位为 0，当数据长度为 2000 时，情况如图 4-17 所示，可以看出，偏移标志此时置 1，2000 字节的传输中，第一片的分片长度为 1480 字节，第二片 500 字节，均表示数据字段，总共 1980 字节，这时加上 IP 首部 20 字节的首部，于是就构成了一个 2000 字节长度的数据报。此外，仔细查看图 4-20，除了分片大小，还可以获取其他分片的个数是 2，而且分别在 408 帧和 409 帧，408 帧为第一帧，数据是 0～1479，409 帧为第二帧，数据为 1480～1979。

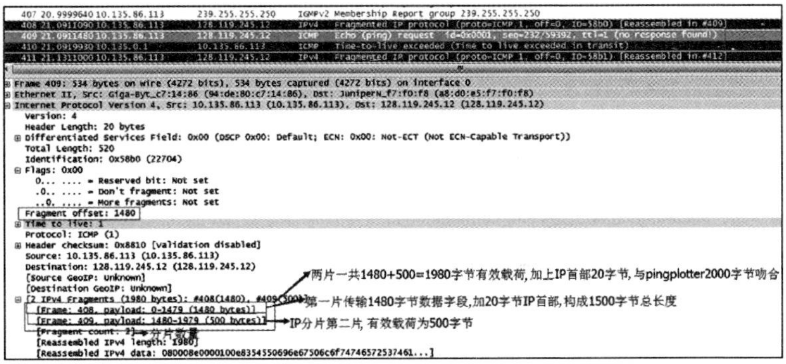

图 4-20　2000 字节数据报分片

用同样的分析方法，来分析当数据大小为 3500 时发生的情况。如图 4-21 所示。具体内容可参考前面数据报长度为 2000 时的分析。

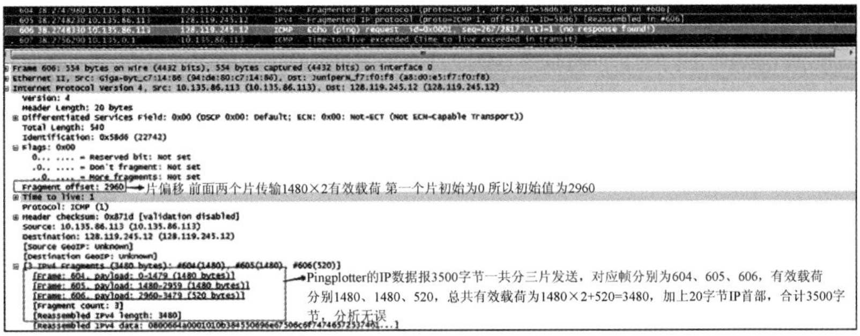

图 4-21　2000 字节数据报分片

如何知道一个报文是否分片呢？查看分片中的标志位 Flag，如图 4-22 所示，当 More Fragment 位为 1 时，就表示该 IP 数据报分片。

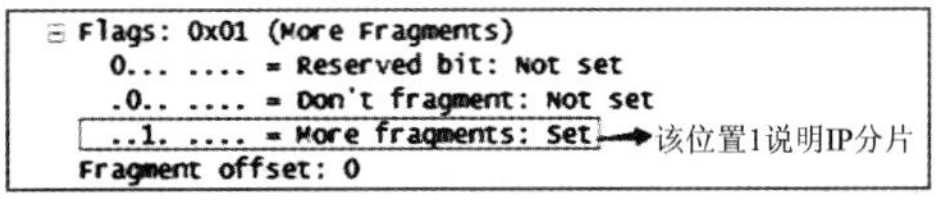

图 4-22　数据报分片的判断标志

4.3 互联网控制消息协议

互联网控制消息协议（ICMP）由 RFC 792 定义，它用于主机和路由器彼此交互网络层信息，ICMP 最典型的用途是差错报告。例如，当运行一次 Telnet、FTP 或 HTTP 会话时，你也许会遇到一些诸如"目的网络不可达"之类的错误报文。这种报文就是在 ICMP 中产生的。

ICMP 通常被认为是 IP 的一部分，但从体系结构上讲它是位于 IP 之上的，因为 ICMP 报文是承载在 IP 分组中的。这就是说，ICMP 报文是作为 IP 有效载荷承载的。类似地，当一台主机收到一个指明上层协议为 ICMP 的 IP 数据报时，它分解该数据报的内容给 ICMP，就像分解一个数据报的内容给 TCP 或 UDP 一样。

4.3.1 ICMP 报文格式及报文类型

1. ICMP 基本格式

IP 首部的 Protocol 值为 1 就说明这是一个 ICMP 报文，如图 4-23 给出了 ICMP 报文的基本格式。

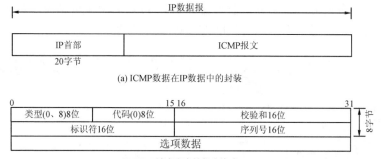

图 4-23　ICMP 报文格式

各种 ICMP 报文的前 4 字节（32 位）都是三个长度固定的字段，即类型字段（8 位）、代码字段（8 位）、校验和字段（16 位）。

8 位类型和 8 位代码字段：一起决定了 ICMP 报文的类型。

16 位校验和字段：包括数据在内的整个 ICMP 数据包的校验和，其计算方法和 IP 头部校验和的计算方法是一样的。

16 位是标识符字段：用于标识本 ICMP 进程。

16 位序列号字段：用于判断回显应答数据报。

ICMP 报文包含在 IP 数据报中，属于 IP 的数据，IP 首部就在 ICMP 报文的前面，一个 ICMP 报文包括 IP 首部（20 字节）、ICMP 首部（8 字节）和 ICMP 报文。

2. ICMP 报文类型

RFC 定义了 13 种 ICMP 报文类型，具体如图 4-24 所示。

类型	类型描述	类型	类型描述
0	响应应答（ECHO-REPLY）	13	时间戳请求
3	不可到达	14	时间戳应答
4	源抑制	15	信息请求（*已作废）
5	重定向	16	信息应答（*已作废）
8	响应请求（ECHO-REQUEST）	17	地址掩码请求
11	超时	18	地址掩码应答
12	参数失灵		

其中类型为 15、16 的信息报文已经作废

图 4-24 ICMP 报文类型

常见的 ICMP 报文如下。

类型 8，代码 0：表示回显请求（ping 请求）。

类型 0，代码 0：表示回显应答（ping 应答）。

类型 11，代码 0：TTL 超时。

4.3.2 ICMP 分析实验

在本节中，将分别通过 ping 和 tracert 两种方法进行数据抓包来对 ICMP 报文结构和内容进行分析，以便更好地学习 ICMP。

1. 实验一：ICMP 和 ping

按如下步骤进行抓包：

（1）打开 Windows 命令提示符（CMD）；

（2）启动 Wireshark，并开始 Wireshark 抓包；

（3）在命令提示符窗口输入"ping-n 10 hostname"或"c：\ windows \ system32 \ ping-n 10 hostname"类型命令，hostname 表示主机名，-n 表示 ping 的次数，默认为 3，在本实验中使用的是"ping-n 10 www.mit.edu"；

（4）当 ping 程序终止时，停止在 Wireshark 抓包。

命令窗口显示的内容如图 4-25 所示。

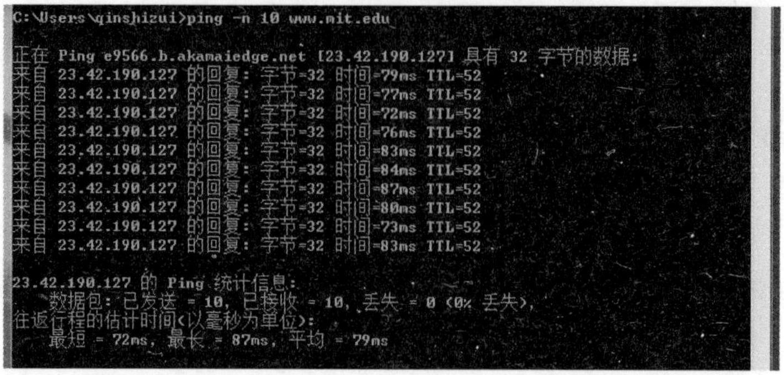

[图片：命令窗口显示 ping -n 10 www.mit.edu 的结果]

图 4-25　命令窗口显示的 ping 内容

Wireshark 抓包显示如图 4-26 所示。

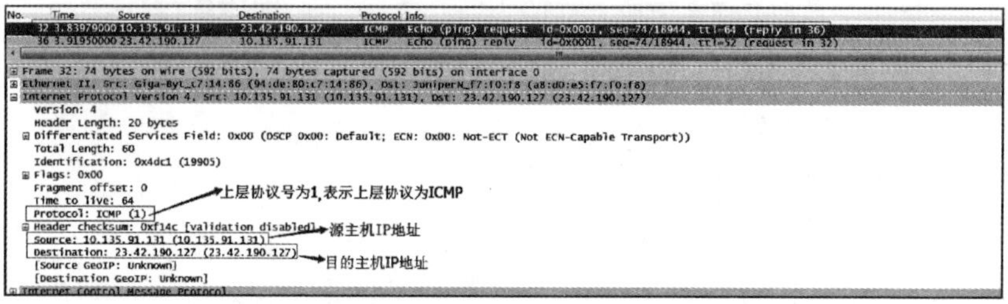

[图片：Wireshark 抓包显示]

图 4-26　Wireshark 抓包显示 ping 信息

通过图 4-26 看出，主机 IP 为 10.135.91.131，目的 IP 是 23.42.190.127，这正是在图 4-24 中显示的 IP 地址。这个地址是从 IP 报文中获得的，很明显，在 ICMP 中没有源和目的地址以及端口号等信息，原因在前面协议分析时提到：ICMP 作为 IP 的上层协议，被封装在 IP 协议中，使用 IP 的首部即可实现数据报在网络中的转发。在 ICMP 报文中可以获得的信息如图 4-27 中所示：ICMP 的校验和信息（0x4d11）、Sequence number、Identifier，并且报文中 Type（类型）是 8，Code（编码）是 0，根据前面的学习，了解到这表示着回显请求，不出意外会在之后看到一个 Type（类型）是 0，Code（编码）是 0 的报文，作为回显回答，如图 4-28 所示。

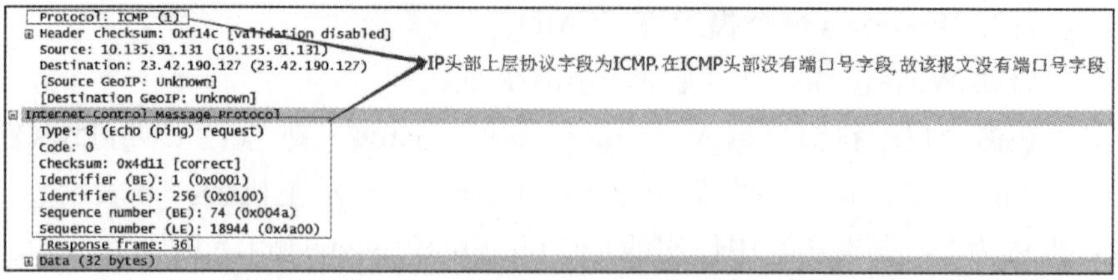

[图片：Wireshark 下 ICMP 报文信息详情]

图 4-27　Wireshark 下 ICMP 报文信息

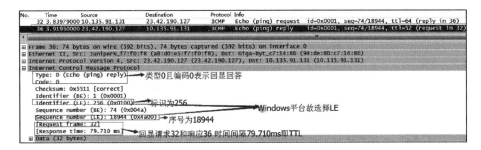

图 4-28　Wireshark 下 ICMP 回显报文信息

2. 实验二：ICMP 和 tracert

按如下步骤进行抓包。

（1）打开 Windows 命令提示符（CMD）。

（2）启动 Wireshark，并开始 Wireshark 抓包。

（3）在命令提示符窗口输入 "tracert hostnameor" 或 "c：\ windows \ system32 \ tracert hostname" 类型命令，hostname 表示主机名，在本实验中使用的是 "tracert www. inria. fr"。

（4）当 tracert 程序终止，停止 Wireshark 抓包。

这时可以看见命令窗口显示的内容，如图 4-29 所示。

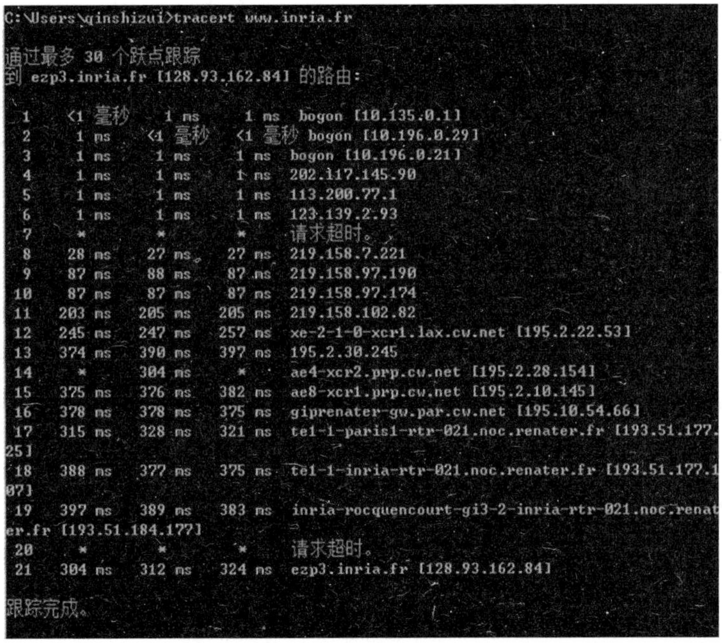

图 4-29　命令窗口显示 tracert 内容

Wireshark 下的报文如图 4-30 所示。

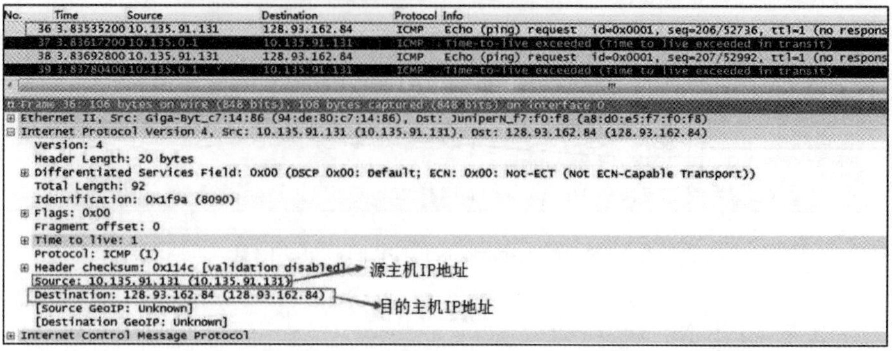

图 4-30　Wireshark 显示 tracert 信息

通过图 4-26 了解到主机 IP 为 10.135.91.131，目的 IP 是 128.93.162.84，正是图 4-30 中 tracert 的 IP 地址。

对比一下 ping 得到的 ICMP 请求报文和 tracert 下 ICMP 的回显报文，如图 4-31 和图 4-32 所示，它们有所不同。在 ping 程序最初的请求报文里面有相应的响应报文，而 tracert 的请求报文没有响应报文。原因是，在命令窗口里面去 ping 一台主机和 tracert 一台主机是不一样的行为，其中 ping 是对目的地进行检验目的主机是否联网，如果在线正常运转就会返回响应报文。而 tracert 是对请求主机通过发送一个个 TTL 逐渐增加的探测包进行路由追踪，在到达目的主机之前，当 TTL 分别为 1、2、3、n 的主机分别达到第 1、2、3、n 跳路由器时返回数据报，在命令窗口呈现的是路由 IP 地址和时间，在 Wireshark 呈现的 error packet 数据包即 TTL-exceeded 报文段，在到达目的主机时候的 request 请求报文就能收到 reply 响应报文。由于 Windows 对路由器跳数进行了限制，即 30 跳，建议使用 pingplotter。

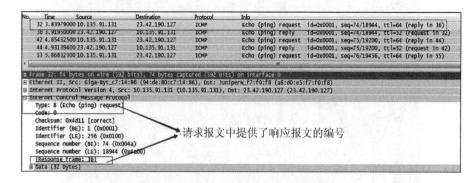

图 4-31　Wireshark 显示 ping 请求信息

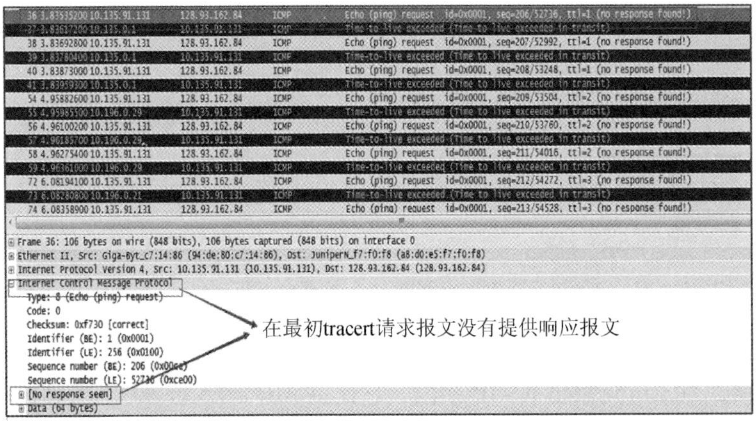

第 4 章 网络层典型协议分析

图 4-32 Wireshark 显示 tracert 回显信息

接下来，再做一个对比，对比 ICMP error packet（出错数据包）和 ICMP echo packet（回显数据包）以及 last three ICMP packets（最后 3 个数据包），这之间有什么区别，如图 4-33 出错数据包、图 4-34 回显数据包、图 4-35 最后 3 个数据包所示。

图 4-33 Wireshark 显示 ICMP 出错数据包（error packet）

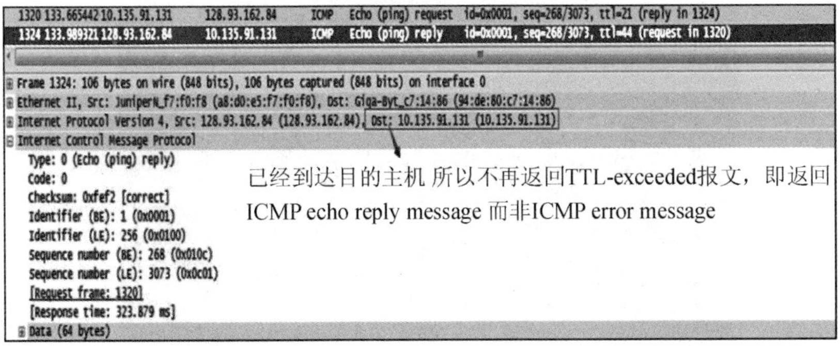

图 4-34 Wireshark 显示 ICMP 回显数据包（echo packet）

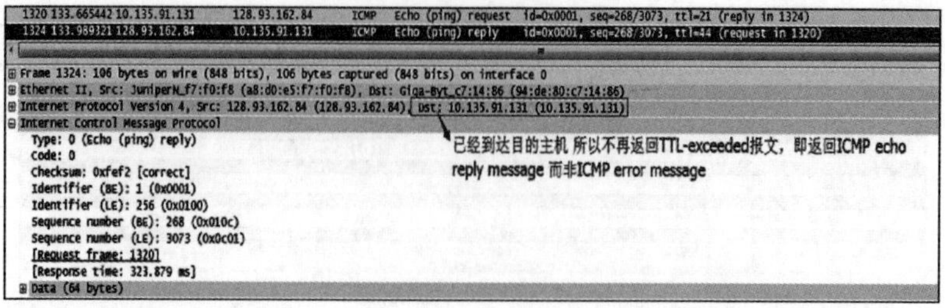

图 4-35　Wireshark 返回 ICMP 最后三个数据包（last three ICMP packets）

注意到 ICMP error packet（出错数据包）有 Type、Code、Checksum、Embedded IPv4 和 Embedded ICMP 字段，它比 ICMP echo packet（回显数据包）多了一个 IPv4 的字段，用于描述第一次请求未到达目的主机，故返回源主机，让它对 TTL 进行加 1 操作，试探下一跳路由器是否是目的主机。其中 ICMP 的 Error packet type 为 11，Code 为 0，但是 ICMP 恢复数据包的 Type 为 0，Code 为 0，含有 Checksum、Identifier、Sequence number 字段，具有额外的时间去响应 ICMP 的回显请求消息。因为 ICMP echo 报在 TTL 过期之前被发送到目的 IP 地址而不是中间路由器。

在进行上述 tracert 追踪实验后，或许会发生一个很特殊的现象，有一个链路连接延迟是明显大于其他路由器的，如图 4-36。之所以会发生这种情况，是因为距离问题，如图 4-37 所示，给出了证明。在实验中一个路由器在中国，而另一个路由器在欧洲。

图 4-36　Tracert 中延迟信息显示

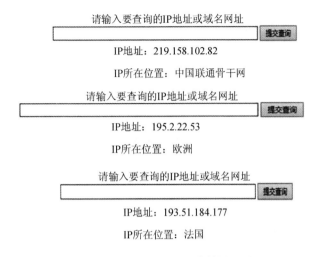

图 4-37 IP 地址所在地查询结果显示

最后需要说明的是，ICMP 发送 UDP 数据包（如 Unix / Linux 平台），那么此时 IP 数据包的上层协议将不再是 01，而应变为 17，表示上层协议是 UDP。

4.4 网络地址转换

进行了有关 Internet 地址和 IPv4 数据报格式讨论后，现在可清楚认识到每个 IP 使能的设备都需要一个 IP 地址。但随着所谓小型办公室、家庭办公室、子网的大量出现，意味着每当一个子网想安装一个局域网以互联多台机器时，都需要 ISP 分配地址供子网中所有机器使用，那么如果网络变大了，当分配了一块地址后还是不够更多的设备使用，应该怎么解决呢？网络地址转换（NAT）技术应运而生。

4.4.1 NAT 简介

如图 4-38 显示了一台 NAT 使能路由器运行的情况，位于家中的 NAT 使能路由器有一个接口，该接口是图 4-38 中右侧所示家庭网络的一部分，家庭网络内的编址中所有 4 个接口都具有相同的网络地址 10.0.0/24。但是，转发到家庭网络之外进入更大的全球 Internet 的分组时显然不能使用这些地址，因为有数以百万计的网络正使用这块地址。这就是说，10.0.0/24 地址仅在给定的网络中才有意义。如何处理地址转换问题，答案在于理解 NAT 的运行过程。

NAT 路由器对外界的行为就如同一个具有单一 IP 地址的单一设备，它对外界隐藏了家庭网络的细节。来自广域网到达 NAT 路由器的所有数据报都具有相同的 IP 地址,这时使用该路由器上的 NAT 转换表,在转换表中找到子网中转换的端口号和 IP 地址。

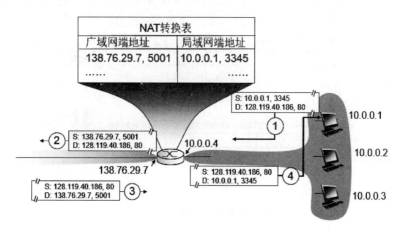

图 4-38　NAT 使能路由器运行情况

继续使用图 4-38,假设一个用户在家庭网络主机 10.0.0.1 旁,请求 IP 地址为 128.119.40.186 的 Web 服务器上的一个 Web 页面。主机 10.0.0.1 为其指派了源端口号 3345 并将该数据报发送到 LAN 中。NAT 路由器收到该数据报,为该数据报生成一个新的源端口号 5001,将源 IP 地址改为其广域网一侧接口的 IP 地址 138.76.29.7(该地址为一个合法可用的公网 IP 地址),且将源端口号 3345 更换为新端口号 5001。当生成一个新的源端口号时,NAT 路由器可选择任意一个当前未在 NAT 转换表中使用的源端口号,路由器的 NAT 也在 NAT 转换表中增加一项。Web 服务器并不知道刚到达的包含 HTTP 请求的数据报已被 NAT 路由器进行了转换,它会发回一个响应数据报,其目的地址是 NAT 路由器的 IP 地址,目的端口号是 5001。当该数据报到达 NAT 路由器时,路由器使用目的 IP 地址与目的端口号从 NAT 转换表中检索出家庭网络浏览器使用的正确 IP 地址(10.0.0.1)和目的端口号(3345)。于是,路由器改写该数据报的目的 IP 地址与目的端口号,并向家庭网络转发该数据报。这就是 NAT 工作的基本过程。

4.4.2　NAT 协议分析实验

NAT 实验不同于之前的 Wireshark 实验,它需要对数据包的输入和输出两方面做出分析,因此需要在两处捕获数据包,在一个 NAT 设备或两台计算机上进行 Wireshark 的抓包分析。为了简化,我们通过在网站 http://gaia.cs.umass.edu/wireshark-labs/wireshark-traces.zip 下载已有的抓包文件,并分析。

在本实验中，在家庭网络中从客户机 PC 向 www.google.com 服务器发送一个请求来进行 NAT 实验分析。图 4-39 显示了 Wireshark 抓包采集场景，Wireshark NAT 实验分为两部分，第一部分是从客户机到 NAT 路由器，抓到的数据包为 NAT_home_side；第二部分是从 NAT 路由器到 ISP 网络，抓到的数据包为 NAT_ISP_side。

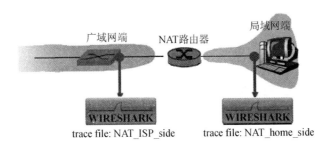

图 4-39 Wireshark 跟踪采集场景

通过 NAT_home_side 和 NAT_ISP_side 两方面对比进行 NAT 协议分析。

1. 对客户机第一条报文分析

① 首先打开 NAT_home_side 报文。找到客户机发出的第一条报文，看到源 IP 地址为 192.168.1.100，如图 4-40 所示。

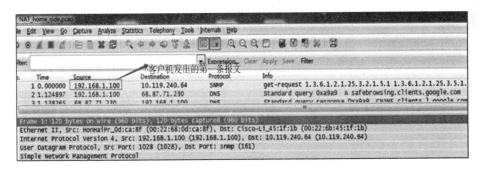

图 4-40 Wireshark 显示 NAT_home_side 报文

② 类比，打开 NAT_ISP_side 报文。找到客户机发出的第一条报文，看到源 IP 地址为 71.192.34.104，如图 4-41 所示。

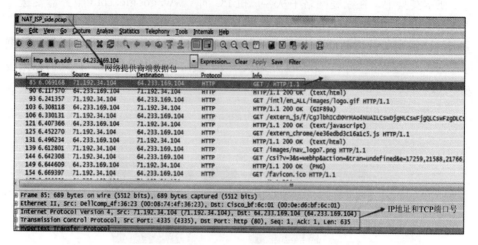

图 4-41 Wireshark 显示 NAT_ISP_side 报文

通过表 4-5,将需要的数据显示如下。

表 4-5 第一条报文 NAT_home_side 和 NAT_ISP_side 数据分析

Packet	Time	Source IP	Destination IP	Source Port	Destination Port
ISP HTTP GET	6.069168	71.192.34.104	64.233.169.101	4335	80
Home HTTP GET	7.109267	192.168.1.100	64.233.169.101	4335	80

2. 对 HTTP 请求过程进行分析

① 在 NAT_home_side 中,HTTP 请求过程如下。

分别截取客户机给谷歌服务器发送的 HTTP GET 请求、200 OK 的回复报文,以及三次握手的 Wireshark 截图,如图 4-42~图 4-44 所示。

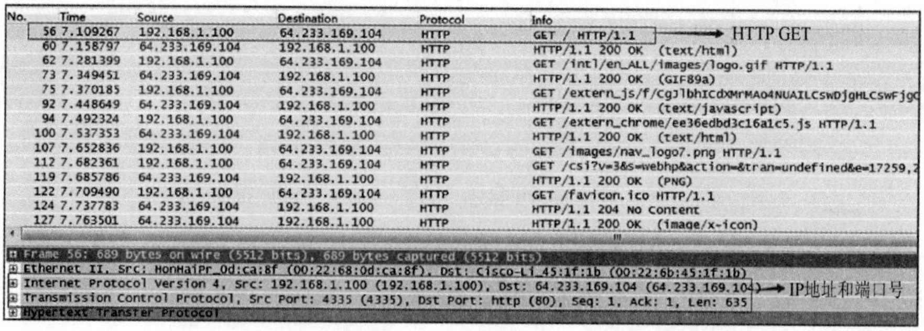

图 4-42 Wireshark 显示 NAT_home_side GET 报文

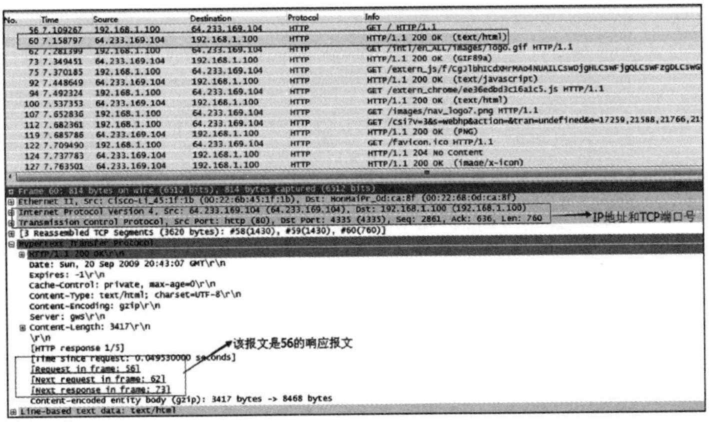

图 4-43 Wireshark 显示 NAT_home_side 200 OK 报文

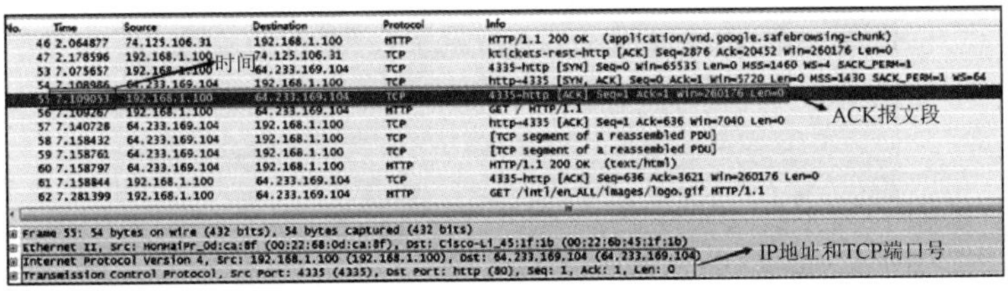

图 4-44 Wireshark 显示 NAT_home_side SYN 和 ACK 报文

② NAT_ISP_side 下，HTTP 请求过程如下。

分别截取客户机给谷歌服务器发送的 HTTP GET 请求、200 OK 的回复报文，以及三次握手的 Wireshark 截图，如图 4-45～图 4-47 所示。

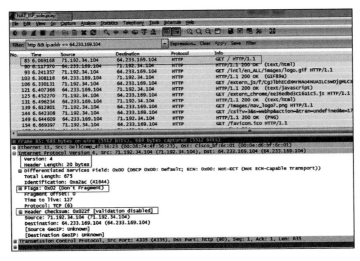

图 4-45 Wireshark 显示 NAT_ISP_side GET 报文

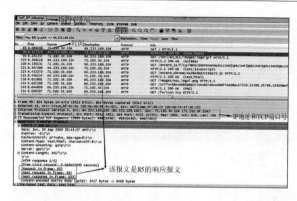

图 4-46 Wireshark 显示 NAT_ISP_side 200 OK 报文

图 4-47 Wireshark 显示 NAT_ISP_side SYN 和 ACK 报文

对上述截图分析，可以了解到表 4-6～表 4-8 的信息。

表 4-6 GET 请求报文 NAT_home_side 和 NAT_ISP_side 数据分析

Packet	Version	Header Length	Flags	Checksum
Home HTTP GET	4	20 bytes	Don't fragment	0xa94a
ISP HTTP GET	4	20 bytes	Don't fragment	0x022f

表 4-7 200 OK 报文 NAT_home_side 和 NAT_ISP_side 数据分析

Packet	Time	Source IP	Destination IP	Source Port	Destination Port
ISP HTTP response	6.117570	64.233.169.101	71.192.34.104	80	4335
Home HTTP response	7.109267	64.233.169.101	192.168.1.100	80	4335

表 4-8 SYN 和 ACK 报文 NAT_home_side 和 NAT_ISP_side 数据分析

Packet	Time	Source IP	Destination IP	Source Port	Destination Port
ISP SYN	6.035475	71.192.34.104	64.233.169.101	4335	80
ISP ACK	6.086754	71.792.34.104	64.233.169.104	4335	80
Home SYN	7.075657	192.168.1.100	64.233.169.101	4335	80
Home ACK	7.109053	192.168.1.100	64.233.169.101	4335	80

NAT 路由器在 SOHO（small office、home office）子网环境中起到转换作用，它收集多个客户机的 IP 地址和 TCP 端口，并赋予新的本地路由器 IP 地址和 TCP 端口，将 IP 数据报发送到 Internet 进行数据传输。

第5章 数据链路层和局域网典型协议分析

数据链路层负责通信节点间在单个链路上的传输活动，实现帧的单跳传输，可以在不可靠的物理链路上实现可靠的传输服务。数据链路层有两种截然不同类型的链路层通道：第一种类型由广播信道组成，这种信道通常在局域网、无线局域网、卫星网和混合光纤电缆（HFC）接入网中；第二种类型的链路层信道是点对点通信链路，如两台路由器之间的通信链路。

本章首先介绍数据链路层的基本服务，接着对数据链路层协议进行介绍，数据链路层协议的例子包括以太网协议（Ethernet）、地址解析协议（ARP）、无线局域网协议 802.11 等。

5.1 数据链路层的概述和服务

在数据链路层的学习中，并不关心一个节点是一台路由器还是一台主机，因此可将路由器和主机均称为节点，并把沿着通信路径连接相邻节点的通信信道称为链路。为了将一个数据报从源主机传输到目的主机，数据报必须通过沿端路径上的每段链路传输，在通过特定的链路时，传输节点将此数据报封装在数据链路层帧中，并将该帧发送到链路上，然后接收节点接收该帧并提取数据报。

数据链路层如图 5-1 所示，该通信路径由一系列通信链路组成，从源主机开始，经过一系列路由器，在目的主机结束。

任何数据链路层的基本服务都是将数据报通过单一通信链路从一个节点移动到相邻节点，但所提供的服务细节会随着数据链路层协议不同而不同。数据链路层协议能够提供的可能服务如下。

成帧：将网络层数据封装到数据链路层帧中，这时数据链路层为了能实现数据有效的差错控制，就采用了一种"帧"的数据块进行传输。而要采取帧格式传输，就必须有相应的帧同步技术，这就是数据链路层的"成帧"（又称"帧同步"）功能。

链路接入：媒体访问协议规定了帧在链路上传输的规则。媒体访问控制（media access control，MAC）协议规定了帧在链路上传输的规则。例如，在链路的一端有一个发送方、链路的另一端有一个接收方的点对点链路，MAC 协议只要链路空闲就可以发送帧。

可靠交付：当数据链路层协议提供可靠交付服务时，它必须保证无差错地经过数据链路层传输每个网络层数据报。

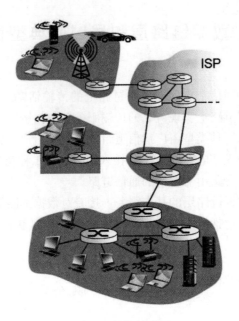

图 5-1　数据链路层

流量控制：流量控制用于防止链路一端的发送节点湮没另一端的接收节点，当没有流量控制时，接收方的缓存区就会溢出，并使帧丢失。

差错检测：网络中最基本的问题是接收方要确信其收到的数据就是数据发送的原始数据而没有错误，为了确保这一点，就需要一种差错检测机制来检测是否发生了差错。传输数据的检测差错的技术有 3 种：奇偶校验、校验和、循环冗余检验（又称 CRC 检验）。

差错纠正：它和差错校测类似，区别在于它不仅检测帧中是否有错误，还能判断哪个比特出现错误。

半双工和全双工：全双工模式下，链路两端可以同时传输，而半双工模式则不可以。

5.2　以太网协议

以太网（Ethernet）是到目前最流行的有线局域网技术，以太网对本地区域联网的重要性就像是 Internet 对全球互联网所具有的地位。

5.2.1　Ethernet 协议报文格式

以太网帧结构如图 5-2 所示。

8字节	6字节	6字节	2字节	46~1500字节	4字节
前同步码	目的地址	源地址	类型	数据	CRC

图 5-2　以太网帧结构

前同步码（8 字节）。设置该字段的目的是指示帧的开始并便于网络中的所有接收器均能与到达帧同步。

目的地址（6 字节）。目的地址字段确定帧的接收者。

源地址（6 字节）。源地址字段标识发送帧的工作站。

类型（2 字节）。2 字节的类型字段仅用于 Ethernet II 帧。该字段用于标识数据字段中包含的高层协议，也就是说，该字段告诉接收设备如何解释数据字段。

数据（46～1500 字节）。如前所述，数据字段的最小长度必须为 46 字节以保证帧长至少为 64 字节，这意味着传输 1 字节信息也必须使用 46 字节的数据字段，如果填入该字段的信息少于 46 字节，该字段的其余部分必须进行填充。

循环冗余检测（cyclic redundancy check，CRC）（4 字节）。目的是使得接收适配器检测帧中是否出现了差错，也就是说，帧中的比特是否发生了翻转。

5.2.2　CSMA/CD

因为以太网能够应用广播，故它需要多路访问协议。以太网使用了广受赞誉的 CSMA/CD 协议。

CSMA/CD 是 carrier sense multiple access with collision detection 的缩写，可译为"载波侦听多路访问/冲突检测"，或"带有冲突检测的载波侦听多路访问"。所谓载波侦听（carrier sense），意思是网络上各个工作站在发送数据前都要侦听总线上有没有数据传输。若有数据传输（称总线为忙），则不发送数据；若无数据传输（称总线为空），立即发送准备好的数据。所谓多路访问（multiple access），意思是网络上所有工作站收发数据共同使用同一条总线，且发送数据是广播式的。所谓冲突（collision），意思是若网上有两个或两个以上工作站同时发送数据，在总线上就会产生信号的混合，两个工作站都辨别不出真正的数据是什么。这种情况称为数据冲突，又称碰撞。为了减少冲突发生后的影响，工作站在发送数据过程中还要不停地检测自己发送的数据，有没有在传输过程中与其他工作站的数据发生冲突，这就是冲突检测（collision detection）。如图 5-3 所示，显示了带冲突检测的 CSMA。

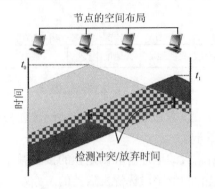

图 5-3 带冲突检测的 CSMA

CSMA/CD 控制方式的优点是原理比较简单，技术上易实现，网络中各工作站处于平等地位，不需集中控制，不提供优先级控制。但在网络负载增大时，发送时间增长，发送效率急剧下降。

CSMA/CD 媒体访问控制方法的工作原理，可以概括为"先听后说，边听边说；一旦冲突，立即停说；等待时机，然后再说"。意思是在发送数据前先侦听信道是否空闲，若空闲，则立即发送数据；若信道忙碌，则等待一段时间至信道中的信息传输结束后再发送数据。若在上一段信息发送结束后，同时有两个或两个以上的节点都提出发送请求，则判定为冲突。若侦听到冲突，则立即停止发送数据，等待一段随机时间，再重新尝试。

上述冲突情况都涉及一个共同算法，即退避算法。

退避算法说明当出现线路冲突时，如果冲突的各站点都采用同样的退避间隔时间，则很容易产生二次、三次的碰撞。因此，要求各个站点的退避间隔时间具有差异性。这要求通过退避算法来实现。

截断的二进制指数退避算法（退避算法之一）如下。

当一个站点发现线路忙时，要等待一个延时时间 M，然后再进行侦听工作。延时时间 M 以以下算法决定。

$M=0 \sim (2^k-1)$ 之间的一个随机数乘以 512 比特时间（例如，对于 10Mbps 以太网，为 $51.2\mu s$），k 为冲突（碰撞）的次数，M 的最大值为 1023，即当 $k=10$ 及以后 M 始终是 $0 \sim 1023$ 之间的一个随机值与 51.2 的乘积，当 k 增加到 16 时，就发出错误信息。

5.2.3 Ethernet 协议分析实验

在这个实验中，将要研究以太网协议，通过数据抓包了解以太网帧格式以及该协议的运行原理。

Ethernet 帧格式分析实验如下。

第 5 章　数据链路层和局域网典型协议分析

抓包分析以太网，按如下步骤抓包：

（1）首先确保你的浏览器的缓存是被清空的，在浏览器资源管理器中，选择工具→Internet 选项→删除文件；

（2）启动 Wireshark，输入下列网址到你的浏览器，即 http://gaia.cs.umass.edu/Wireshark-labs/HTTP-ethereal-lab-file3.html；

（3）停止抓包，首先找到你的主机发送到 gaia.cs.umass.edu 的 HTTP GET 消息，同时找到 HTTP Response 消息；

（4）在窗口点击 Analyze→Enabled Protocols，取消 IP 框并选择 OK。

Wireshark 将显示如图 5-4 的分析窗口，选择包含 HTTP 消息的以太网帧，并进行协议分析。

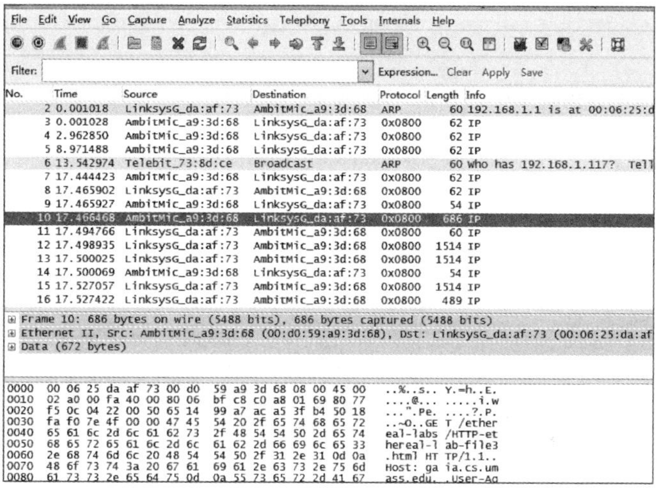

图 5-4　Wireshark 显示以太网分析窗口

如图 5-5 所示，可以看到，在 HTTP GET 消息中，源 MAC 地址为 94:de:80:c7:14:86，目的 MAC 地址为 a8:d0:e5:f7:f0:f8。

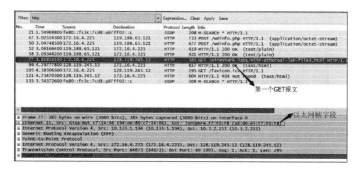

图 5-5　Wireshark HTTP GET 消息中的 MAC 地址分析

如图 5-6 所示，在响应报文中，以太网源地址是 a8∶d0∶e5∶f7∶f0∶f8，既不是主机的 MAC 地址，也不是网站主机的 MAC 地址，而是和主机相连的第一个路由器的适配器 MAC 地址。以太网帧的目的地址是 94∶de∶80∶c7∶14∶86，它是主机的 MAC 地址。

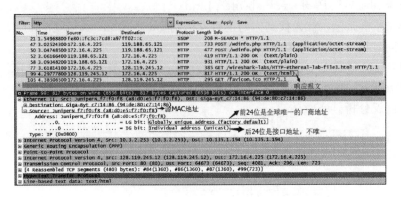

图 5-6　Wireshark 以太网地址分析

数据链路层的上层协议对应的类型字段的值为 0x080，表示上层协议是 IP，即以太网封装 IP 协议，如图 5-7 所示。

图 5-7　数据链路层上层协议字段

当对数据报 Frame 进行分析时，在包含 "GET" 的以太网帧中，十六进制数表示的字符 "G" 的位置从第 55 字节开始，G 出现前总共有 54 字节，其中以太网帧有 14 字节，IP 首部有 20 字节，TCP 首部有 20 字节，和为 54 字节，如图 5-8 所示。

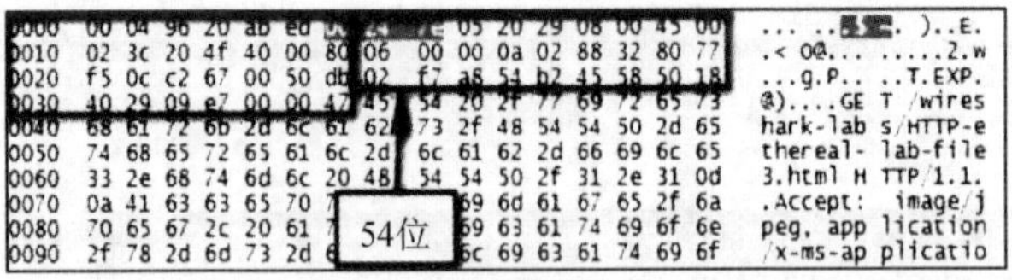

图 5-8　Wireshark 显示以太网帧

但是，在以太网帧中是不具有 200 OK 的字段消息的，在帧中不会显示出字母 "O"。

5.3 地址解析协议

地址解析协议（ARP）是对网络层地址（如 Internet 的 IP 地址）和数据链路层地址（即 MAC 地址）进行转换的协议，学习 ARP 之前，首先介绍一下数据链路层地址，即 MAC 地址。

5.3.1 MAC 地址

首先需要说明，并不是节点（路由器或主机）具有 MAC 地址，而是节点的适配器具有数据链路层地址。如图 5-9 所示，说明了这种情况。局域网地址有各种不同的称呼：LAN 地址（LAN address）、物理地址（physics address）或 MAC 地址（MAC address）。

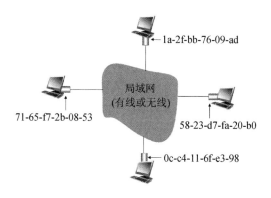

图 5-9 与局域网相连的每个适配器都有一个唯一的 MAC 地址

IP 地址与 MAC 地址在计算机里都是以二进制表示的，IP 地址是 32 位，而 MAC 地址则是 48 位。MAC 地址的长度为 48 位（6 字节），通常表示为 12 个十六进制数，每 2 个十六进制数之间用冒号隔开，如 08:00:20:0a:8c:6d 就是一个 MAC 地址，其中前 6 位十六进制数 08:00:20 代表网络硬件制造商的编号，它由 IEEE（电气与电子工程师协会）分配，而后 3 位十六进制数 0a:8c:6d 代表该制造商所制造的某个网络产品（如网卡）的系列号。只要不去更改自己的 MAC 地址，那么 MAC 地址在世界是唯一的。

MAC 地址应用于局域网和广域网中的计算机之间的通信。在局域网中由于网络的结构相对简单一些，很少涉及三层交换机（也就是路由器），所以可以用来标识每台主机的就只有 MAC 地址，这种情况下交换机的每一个端口对应的就是一个 MAC 地址。当交换机收到数据包之后与自己的 MAC 表项进行比对，如果发现源主机和目的主机并不在同一个端口，但 MAC 表项中有对应的 MAC 地址时，它就按照 MAC 地址表对数据包进行转发；如果 MAC 表项中没有对应的目的地址，交换机就对所有的端口进行广播（除源端口）。

5.3.2 ARP 简介

地址解析协议（address resolution protocol，ARP）是用于将 IP 地址映射为 MAC 地址的网间协议。所谓"地址解析"就是主机在发送帧前将目的 IP 地址转换成目的 MAC 地址的过程。ARP 的基本功能就是通过目的设备的 IP 地址，查询目的设备的 MAC 地址，以保证通信顺利进行。在每台安装有 TCP/IP 协议的计算机里都有一个 ARP 缓存表，表里的 IP 地址与 MAC 地址是一一对应的，如表 5-1 所示。

表 5-1 ARP 缓存表

IP 地址	MAC 地址	TTL
192.168.1.100	08：00：20：0a：8c：6d	13：45：00
192.168.1.200	08：00：20：0a：8c：7a	13：52：00
192.168.1.203	00：aa：00：62：c6：09	13：56：00
⋮	⋮	⋮

以主机 A（192.168.1.5）向主机 B（192.168.1.1）发送数据为例。当发送数据时，主机 A 会在自己的 ARP 缓存表中寻找是否有目的 IP 地址。如果找到了，也就知道了目的 MAC 地址，直接把目的 MAC 地址写入帧里面发送；如果在 ARP 缓存表中没有找到相对应的 IP 地址，主机 A 就会在网络上发送一个广播，广播的目的 MAC 地址为"ff:ff:ff:ff:ff:ff"，这表示向同一网段内的所有主机发出这样的询问，即"192.168.1.1 的 MAC 地址是什么？"网络上其他主机并不响应 ARP 询问，只有主机 B 接收到这个帧时，才向主机 A 做出这样的回应，即"192.168.1.1 的 MAC 地址是 00：aa：00：62：c6：09"。这样，主机 A 就知道了主机 B 的 MAC 地址，它就可以向主机 B 发送信息了。同时它还更新了自己的 ARP 缓存表，下次再向主机 B 发送信息时，直接从 ARP 缓存表里查找。ARP 缓存表采用了老化机制，在一段时间内如果表中的某一行没有使用，就会被删除，这样可以大大减少 ARP 缓存表的长度，加快查询速度。

ARP 缓存表是可以查看的，也可以添加和修改。在命令提示符下，输入"arp-a"查看 ARP 缓存表中的内容。用"arp-d"命令可以删除 ARP 表中某一行的内容；用"arp-s"可以手动在 ARP 表中指定 IP 地址与 MAC 地址的对应。图 5-10 显示了 arp-a 命令。

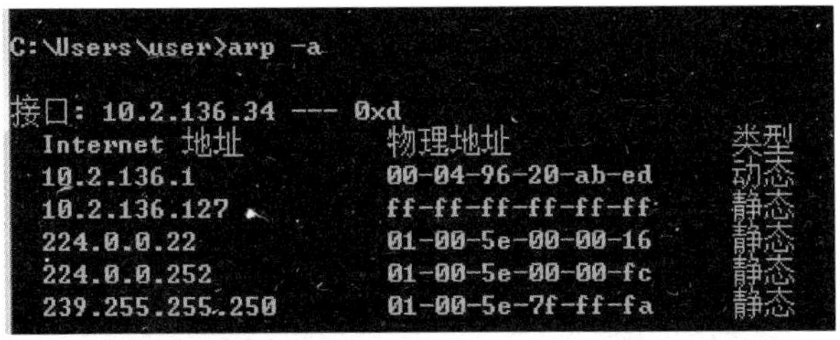

图 5-10 命令窗口显示 arp-a 命令。

5.3.3 ARP 分析实验

在这个实验中，我们研究 ARP 如何运行。

利用 MS－DOS 命令 arp 或 c:\windows\system32\arp 查看主机上 ARP 缓存的内容。

按如下步骤抓包。

（1）利用 MS-DOS 命令 arp-d * 清除主机上 ARP 缓存的内容。

（2）清除浏览器缓存。

（3）启动 Wireshark 并开始抓包。

（4）在浏览器的地址栏中输入 http：//gaia.cs.umass.edu/Wireshark-labs/HTTP-Wireshark-lab-file3.html。

（5）停止分组俘获，选择"Analyze→Enabled Protocols"，取消对 IP 复选框的选择，单击 OK，窗口如图 5-11 所示。

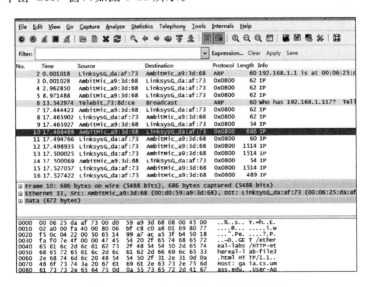

图 5-11 Wireshark 显示 ARP 抓包数据

用 arp-a 查看主机内容，如图 5-12 所示，其中第一列表示 Internet 地址（IP 地址）、第二列表示物理地址（MAC 地址），第三列表示类型。ARP 缓存表表示在主机适配器已经存储的 IP 地址及其对应的 MAC 地址。

图 5-12　命令窗口显示主机缓存信息

需要说明的是，在 Wireshark 捕获到的以太网帧中不存在 CRC 字段。通过对 Frame 首部Type字段的十六进制值分析时，以太网帧类型字段的值是 0x0806，表示上层协议是 ARP，即地址解析协议，如图 5-13 所示。

图 5-13　Wireshark 显示以太网帧 Type 字段

如图 5-14 所示，显示了对一个 ARP 请求报文进行分析。编码字段的值为 1，表示 ARP 请求。ARP 报文段里包含的发送方 IP 地址是 10.135.1.194，由于目的 MAC 地址未知，所以目的 MAC 地址暂时是 00:00:00:00:00:00。

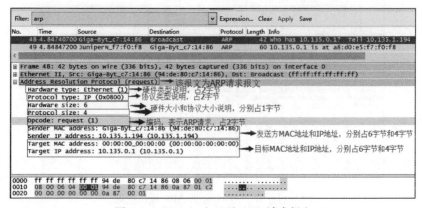

图 5-14　Wireshark 显示 ARP 请求报文

如图 5-15 所示，从 Frame 起始字节的位置算起，opcode 字段是从第 21 字节开始的。

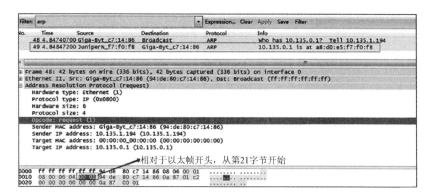

图 5-15　Wireshark 显示 ARP 请求报文 opcode 字段

接着，如图 5-16 所示，在 ARP 响应报文看到发送方 MAC 地址为 a8：d0：e5：f7：f0：f8，源 IP 地址为 10.135.0.1，目的 MAC 地址为 94：de：80：c7：14：86，目的 IP 为 10.135.1.194。ARP opcode field 的值是 2，表示该报文为响应报文。在响应报文中 ARP opcode field 字段从第 21 字节开始，距离以太帧第一个字节是 20 字节。

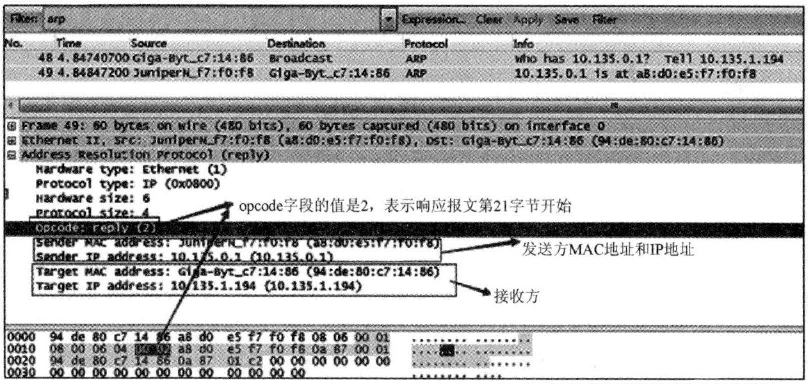

图 5-16　Wireshark 显示 ARP 响应报文

5.4　无线局域网协议 802.11

当前，无线局域网在工作场所、家庭、教育机构、咖啡馆、机场以及街头得到了广泛的使用，它已成为一种最为常用的 Internet 接入手段。本节将分析学习无线局域网协议 IEEE 802.11，又称 Wi-Fi。

5.4.1 802.11 简介

有关无线局域网的 802.11 标准体系包括 802.11b、802.11a、802.11g、802.11n 以及 802.11ac 等。表 5-2 中总结了这些标准的主要特征。

表 5-2 IEEE 802.11 标准主要特征

标准	频率范围	数据率
802.11b	2.4~2.485 GHz	最高为 11Mbps
802.11a	5.1~5.8 GHz	最高为 54Mbps
802.11g	2.4~2.485 GHz	最高为 54Mbps
802.11n	2.4GHz，5GHz	最高为 600Mbps
802.11ac	5GHz	最高为 1Gbps

这五个 802.11 标准具有许多共同的特征，目前最新标准为 802.11ac。它们都使用了相同的媒体访问协议 CSMA/CA，数据链路层都使用了相同的帧格式，都具有降低传输速率以到达更远距离的能力。

IEEE 802.11 体系结构如下。

如图 5-17 所示，显示了 802.11 无线局域网体系结构的基本构件。802.11 体系结构的基本构件是基本服务集（BSS），一个 BBS 通常包含一个或多个无线站点和一个在 802.11 术语中称为接入点（AP）的中央基站。

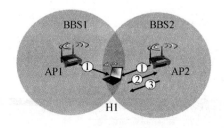

图 5-17 IEEE 802.11 体系结构

与以太网设备类似，每个 802.11 无线站点都具有一个 6 字节的 MAC 地址，该地址存储在该站适配器的固件中。每个 AP 的无线接口也具有一个 MAC 地址，这些 MAC 地址由 IEEE 管理，理论上是全球唯一的。

802.11 定义了两种类型的设备，一种是无线站，通常是通过一台 PC 加上一块无线网络接口卡构成的，另一个称为无线接入点（access point，AP），它的作用是提供无线和有线网络之间的桥接。

IEEE 802.11 帧格式如下。

如图 5-18 所示，给出了一般 802.11 帧格式。

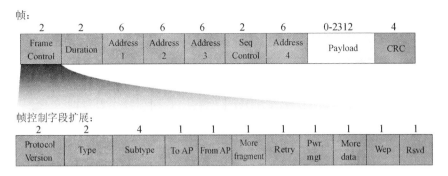

图 5-18 802.11 帧格式

下面给出部分字段的介绍。

MAC Header（MAC 头）：Frame Control（帧控制域）、持续时间/标识（Duration/ID）、地址域（Address）、序列控制（Seq Control）、有效载荷（Payload）、CRC。

Frame Body（帧体部分）：包含信息根据帧的类型有所不同，主要封装的是上层的数据单元，长度为 0～2312 字节，可以推出，802.11 帧最大长度为 2346 字节。

Subtype：指明数据帧的子类型，因为即便是控制帧，还分 RTS 帧、CTS 帧、ACK 帧等，通过这个域判断出该数据帧的具体类型。

Protocol Version（协议版本）：显示该帧所使用的 MAC 版本，通常为 0。

Type（类型域）和 Subtype（子类型域）：共同指出帧的类型。

More fragment：分片标志，若数据帧被分片了，那么这个标志为 1，否则为 0。

Retry：表明是否是重发的帧，若是为 1，不是为 0。

Pwr. mgt：当网络主机处于省电模式时，该标志为 1，否则为 0。

More data：当 AP 缓存了处于省电模式下的网络主机的数据包时，AP 给该省电模式下的网络主机的数据帧中该位为 1，否则为 0。

Wep：加密标志，若为 1 表示数据内容加密，否则为 0。

802.11 帧主要有三种类型，数据帧负责在工作站之间搬运数据；控制帧负责区域的清空、信道的取得以及载波监听的维护，并于收到数据时予以肯定确认，借此提高工作站之间数据传送的可靠性；管理帧负责监督，主要用来加入或退出无线网络以及处理 AP 之间关联的转移事宜。如图 5-19 所示，给出了三种帧的格式。

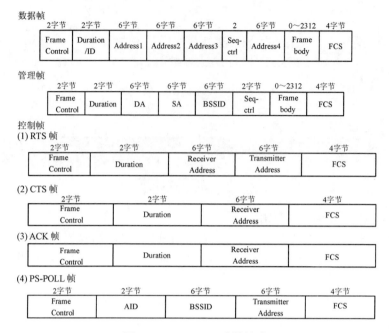

图 5-19 802.11 三种帧格式

5.4.2 CSMA/CA 访问协议

802.11 的设计者为 802.11 无线局域网选择了一种随机访问协议，称为带碰撞避免的 CSMA，CSMA/CA 利用 ACK 信号来避免冲突的发生，也就是说，只有当客户端收到网络上返回的 ACK 信号后才确认送出的数据已经正确到达目的地址。

为了尽量减少碰撞，802.11 标准采用了一种称为虚拟载波监听地的机制，就是让源站把它要占用的信道时间（包括目的站发回确认帧所需时间）写入所发送的数据帧中，以便使其他所有的站在这一段时间都不要发送数据。"虚拟载波监听"的意思是其他各站并没有监听信道，而是由于这些站都知道了源站正在用信道才不发送数据。信道处于忙状态就表示由于物理层的载波监听检测到信道忙，或者由于 MAC 层的虚拟载波监听机制指出了信道忙。

SIFS，即短(short)帧间间隔。SIFT 是最短的帧间间隔，用来分隔属于一次对话的各帧。使用 SIFS 的帧类型有 ACK 帧、CTS 帧、由过长的 MAC 帧分片后的数据帧，以及所有回答 AP 探询的帧和在 PCF 方式中 AP 发送出的任何帧。

CSMA/CA 工作原理如图 5-20 所示。

（1）先检测信道。若检测到信道空闲，则等待一段时间 DIFS 后就发送整个数据帧，并等待确认。

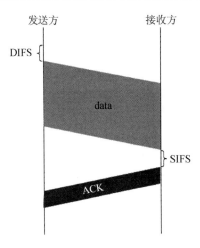

图 5-20　CSMA/CA 工作原理

（2）目的站若正确收到此帧，则经过时间间隔 SIFS 后，向源站发送确认帧 ACK。

（3）其他站都设置网络分配向量 NAV，表明在这段时间内信道忙，不能发送数据。

（4）当确认帧 ACK 结束时，NAV 也就结束了，在经历了帧间间隔之后，接着会出现一段空闲时间，称为争用窗口，表示在这段时间内有可能出现各站点争用信道的情况。

5.4.3　802.11 协议分析实验

在之前的实验中，我们都是用有线以太网进行的。本次实验的 820.11 协议是无线网络协议，我们抓包的数据帧是"空气中的"，而普通的无线网卡不具备抓包功能。所以，建议用现有的数据包进行分析，下载地址为 http://gaia.cs.umass.edu/wireshark-labs/wireshark-traces.zip。

802.11 实验中，主要对数据帧、控制帧、管理帧三种帧的格式，通过 Wireshark 实验进行分析。

1. 数据帧

Wireshark 显示的数据帧格式如图 5-21 所示。

在图 5-21 中，可以看出 Version、Type 和 Subtype 的 08H，即 00001000，后两位 00，表明协议版本为 0，倒数 3、4 位 10 表明这是一个数据帧，前四位 0000 是 Subtype。

Frame Control 后 8 位 0AH，即 00001010，To DS=0，From DS=1，表明该数据帧来自 AP。More fragment=0，表明这是该帧的最后一段，Retry=1，

表明这是重传帧,Pwr. mgt.＝0,表明发送方没有进入节能模式;More data＝0 表明没有更多的帧,即 No data buffered。Protected＝0,表明没有加密,Order＝0, 表明没有严格的顺序要求。

图 5-21　Wireshark 显示数据帧

Duration 位为 d500,高位为 00,低位为 d5,所以持续时间为 00d5H＝ 213μs。Address 1＝0022698ea744,接收方的 MAC 地址;Address2＝0611b51a0a05, 发送发地址,即 AP 地址;Address 3 ＝ 00005e00040a,远程远端地址; Sequence＝3032,低位为 32,高位为 30,即 0011001000110010,段号为 0,帧号为 0011 0010 0011B＝803D,check sequence＝23093131H,检测结果正确。

在图 5-22 中,看到 Sequence 中的 12 位帧号递增,Sequence 为 804,可看出这是图 5-21 的下一帧数据帧,Address 1 ＝0022698ea744,即接收方的 MAC 地址;Address 2＝0611b51a0a05,发送方地址,即 AP 地址;Address 3＝00005e00040a, 即远程远端地址。这三个地址与图 5-21 中的帧一致,是同一发送方发送给同一接收方的连续两帧,帧号＝804,帧号递增。

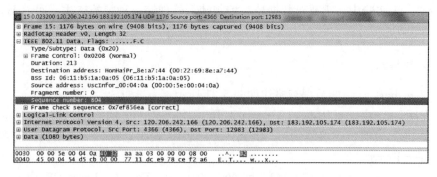

图 5-22　Wireshark 显示数据帧

2. 控制帧

(1) RTS 帧如图 5-23 所示。

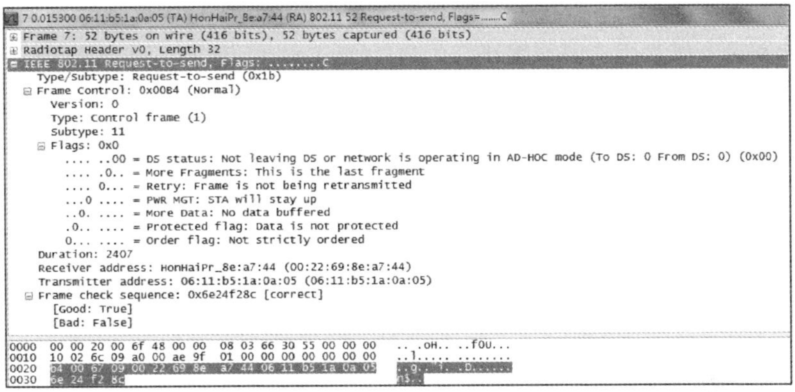

图 5-23　Wireshark 显示 RTS 帧

通过图 5-23，可看出 Version、Type 和 Subtype 的 b4H，即 10110100，后两位 00 表明协议版本为 0，倒数 3、4 位 01 表明这是一个控制帧，前四位 1011 是 Subtype，表明这是 RTS 帧。

Frame Control 后 8 位 00H，控制帧的这几位除 Pwr.mgt 外必然是 0。Pwr.mgt，即发送方没有进入节能模式。

Duration 位为 6709，高位为 67，低位为 09，所以持续时间为 096fH = 2407μs。Receiver Address = 00:22:69:8e:a7:44，即接收方的 MAC 地址；Transmitter Address = 06:11:b5:1a:0a:05，即发送方地址；check sequence = 6e24f28cH，检测结果正确。

(2) CTS 帧如图 5-24 所示。

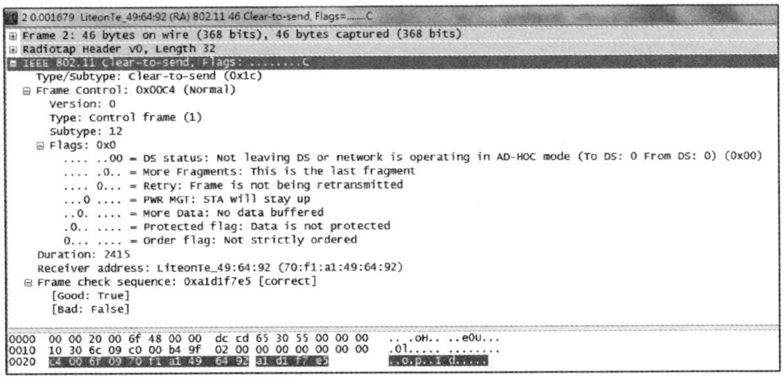

图 5-24　Wireshark 显示 CTS 帧

在图 5-24 中，可看出 Version、Type 和 Subtype 的 C4H，即 11000100，后两位 00，表明协议版本为 0，倒数 3、4 位 01 表明这是一个控制帧，前四位 1100 是 Subtype，表明这是一个 CTS 帧。

Frame Control 后 8 位 00H，控制帧的这几位除 Pwr.mgt 外必然是 0。Pmr.mgt，即发送方没有进入节能模式。

Duration 位为 6f09，高位为 6f，低位为 09，所以持续时间为 096fH＝$2415\mu s$。Receiver Address＝70f1a1496492，即接收方的 MAC 地址；check sequence＝a1d1f7e5H，检测结果正确。

(3) ACK 帧如图 5-25 所示。

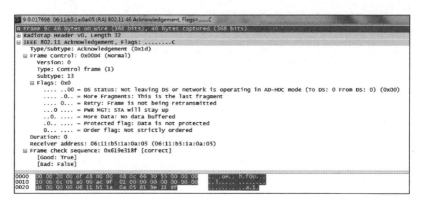

图 5-25　Wireshark 显示 ACK 帧

在图 5-25 中，可看出 Version、Type 和 Subtype 的 d4H，即 11010100，后两位 00，表明协议版本为 0，倒数 3、4 位 01 表明这是一个控制帧，前四位 1101 是 Subtype，表明这是一个 ACK 帧。

Frame Control 后 8 位 00H，控制帧的这几位除 Pwr.mgt 必然是 0。Pwr.mgt，即发送方没有进入节能模式。

Duration 位为 0000，所以持续时间为 0，ACK 表明该帧的传送结束，所以持续时间为 0，Receiver Address＝00：22：69：8e：a7：44，即接收方的 MAC 地址；check sequence＝6e24f28cH，检测结果正确。

(4) Block ACK 帧如图 5-26 所示。

在图 5-26 中，可看出 Version、Type 和 Subtype 的 94H，即 10010100，后两位 00，表明协议版本为 0，倒数 3、4 位 01 表明这是一个控制帧，前四位 1001 是 Subtype，表明这是一个 Block ACK，一个块确定帧。

Frame Control 后 8 位 00H，控制帧的这几位除 Pwr.mgt 必然是 0。Pwr.mgt，即发送方没有进入节能模式。

Duration 位为 9400，高位为 94，低位为 00，所以持续时间为 0094H＝

148μs。Receiver Address ＝70f1al496492，即接收方的 MAC 地址。Transmitter Address ＝3822d67704d3，即发送方的 MAC 地址；check sequence＝d2ed060f，检测结果正确。其余字段如下：Block Ack Type＝02H，即 Compressed Block；Block Ack Control＝0005H，Block Ack Starting Sequence Control＝9320H。

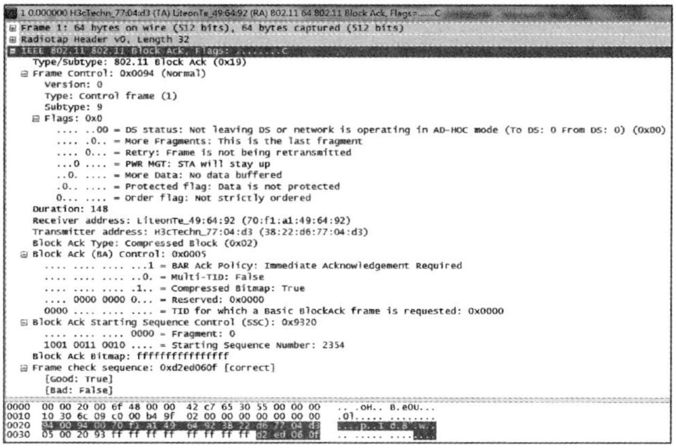

图 5-26 Wireshark 显示 Block ACK 帧

3. 管理帧

Wireshark 显示的管理帧如图 5-27 所示。

在图 5-27 中，可看出 Version、Type 和 Subtype 的 80H，即 1000 0000，后两位 00，表明协议版本为 0，倒数 3、4 位 00 表明这是一个管理帧，前四位 1000 是 Subtype，表明这是信标帧，AP 每隔一段时间就会发出 Beacon（信标）信号，用来宣布 802.11 网络的存在。

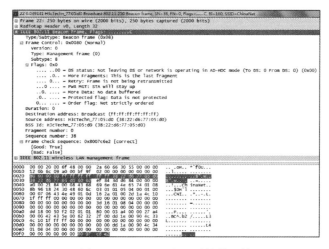

图 5-27 Wireshark 显示的管理帧

Frame Control 后 8 位为 00H，To DS＝0，From DS＝0，管理帧这两位固定。More fragments＝0，表明这是该帧的最后一段；Retry＝0，表明这不是重传帧，Pwr.mgt＝0，表明发送方没有进入节能模式，More Data＝0 表明没有更多的帧，即 No data buffered，该位被置是因为有 AP 给在休眠中的主机缓存了数据，由于 To DS＝0，From DS＝0，数据在主机之间传送，所以 More Data 必定为 0. Protected＝0，表明没有加密，Order＝0，表明没有严格的顺序要求。

Duration 位为 0000，信标帧传送完，此次传输就已经结束，所以持续时间为 0。Destination Address＝ff ff ff ff，即广播；Source Address＝3822d67705d0，为 AP 地址；BSSID＝3822d67705d0，即 AP 地址；Sequence＝6002，低位为 02，高位为 60，即 0000 0010 0110 0000，段号为 0，帧号为 0000 0010 0110B＝38D，check sequence＝8007c6e2H，检测结果为正确。

第二部分

套接字编程实践

第 6 章 TCP 和 UDP 套接字编程

6.1 什么是套接字

套接字，即 socket，是一种通信机制，凭借这种机制，客户机/服务器（即要进行通信的进程）系统的开发工作既可以在本地单机上进行，也可以跨网络进行。也就是说，它可以让不在同一台计算机但通过网络连接计算机上的进程进行通信。也因为这样，套接字明确地将客户端和服务器区分开来。

6.2 套接字的属性

套接字的特性由 3 个属性确定，它们分别是域、类型和地址。

1. 域

它指定套接字通信中使用的网络介质，最常见的套接字域是 AF_INET，它指的是 Internet 网络。当客户使用套接字进行跨网络连接时，它就需要用到服务器计算机的 IP 地址和端口来指定一台联网机器上的某个特定服务，所以在使用套接字作为通信的终点时，服务器应用程序必须在开始通信之前绑定一个端口，服务器在指定的端口等待客户的连接。另一个域 AF_UNIX 表示 Unix 文件系统，它就是文件输入/输出，而它的地址就是文件名。

2. 类型

Internet 提供了两种通信机制，即流(stream)和数据报(datagram)，因而套接字的类型分为流套接字和数据报套接字。

流套接字由类型 SOCK_STREAM 指定，它们是在 AF_INET 域中通过 TCP 连接实现的，同时也是 AF_UNIX 中常用的套接字类型。流套接字提供的是一个有序、可靠、双向字节流的连接，因此发送的数据可以确保不会丢失、重复或乱序到达，而且它还有一定的出错后重新发送的机制。

与流套接字相对的是由类型 SOCK_DGRAM 指定的数据报套接字，它不需要建立连接和维持一个连接，它们在 AF_INET 中通常是通过 UDP 协议实现的。它对可以发送的数据的长度有限制，数据报作为一个单独的网络消息被传输，它可能会丢失、复制或错乱到达，UDP 不是一个可靠的协议，但是它的速度比较高,因为它并不需要总是建立和维持一个连接。

3. 地址

每个套接字都有其自己的地址格式,对于 AF_UNIX 域套接字,它的地址由结构 sockaddr_un 来描述,该结构定义在头文件 sys/un.h 中,它的定义如下:

```
struct sockaddr_un {
    sa_family_t sun_family;    //AF_UNIX,它是一个短整型
    char sum_path [];    //路径名
};
```

对于 AF_INET 域套接字,它的地址结构由 sockaddr_in 来描述,它至少包括以下几个成员。

```
struct sockaddr_in {
    short int    sin_family; //AF_INET
    unsigned short int    sin_port;    //端口号
    struct in_addr    sin_addr;    //IP 地址
};
```

而 in_addr 定义如下。

```
struct in_addr {
    unsigned long int s_addr;
};
```

6.3 服务器端与客户端

1. 服务器端

首先服务器应用程序用系统调用 socket 来创建一个套接字,它是系统分配给该服务器进程的类似文件描述符的资源,它不能与其他的进程共享。接下来,服务器进程会给套接字起个名字,我们使用系统调用 bind 来给套接字命名。然后服务器进程就开始等待客户连接到这个套接字。然后,系统调用 listen 来创建一个队列并将其用于存放来自客户的进入连接。最后,服务器通过系统调用 accept 来接受客户的连接。它会创建一个与原有的命名套接字不同的新套接字,这个套接字只用于与这个特定客户端进行通信,而命名套接字(即原先的套接字)则被保留下来继续处理来自其他客户的连接。

2. 客户端

基于 socket 的客户端比服务器端简单,同样,客户应用程序首先调用 socket

来创建一个未命名的套接字,然后将服务器的命名套接字作为一个地址来调用 connect 与服务器建立连接。一旦连接建立,我们就可以像使用底层的文件描述符那样用套接字来实现双向数据的通信。

6.4 运输层套接字的使用

6.4.1 面向连接的 TCP 套接字编程

1. 服务器端流程

①创建套接字(socket);②将套接字绑定到一个本地地址和端口上(bind);③将套接字设定为监听模式,准备接受客户端请求(listen);④等待客户端请求到来,当请求到来后,接受连接请求,返回一个新的对应于此连接的套接字(accept);⑤用返回的套接字和客户端进行通信(send/recv);⑥返回,等待另一个客户端请求;⑦关闭套接字。

2. 客户端流程

①创建套接字(socket);②向服务器发出连接请求(connect)和服务器进行通信(send/recv);③关闭套接字。

6.4.2 无连接的 UDP 套接字编程

1. 服务器端流程(接收端)

①创建套接字(socket);②将套接字绑定到一个本地地址和端口上(bind);③用返回的套接字和客户端进行通信(recvfrom);④关闭套接字。

2. 客户端流程(发送端)

①创建套接字(socket);②向服务器发送数据(sendto);③关闭套接字。

6.5 Windows 平台 TCP 套接字的接口及使用

6.5.1 套接字流程

1. 创建套接字——socket 系统调用

socket 函数用于创建一个套接字。
语法为

SOCKET socket (int af, int type, int protocol);

af：标识一个地址家族，通常为 AF_INET。

type：标识套接字类型，如果为 SOCK_STREAM，表示流套接字；如果为 SOCK_DGRAM，表示数据报套接字。

protocol：标识一个特殊的地址被用于这个套接字，通常为 0，表示采用默认的 TCP/IP 地址。

2. bind 函数绑定套接字

bind 函数用于将套接字绑定到一个已知的地址上。如果函数执行成功，返回值为 0，否则为 SOCKET_ERROR。

语法为

int bind (SOCKET s, const struct sockaddr FAR *name, int namelen);

s：一个套接字。

name：一个 sockaddr 结构指针，该结构中包含了要绑定的地址和端口号。

namelen：确定 name 缓冲区的长度。

在定义一个套接字后，需要调用 bind 函数为其指定本机地址、地址和端口号。

3. listen 函数监听

listen 函数用于将套接字置入监听模式。

语法为

int listen (SOCKET s, int backlog);

s：套接字。

backlog：表示等待连接的最大队列长度。例如，如果 backlog 被设置为 3，此时有 4 个客户端同时发出连接请求，那么前 3 个客户端连接会放置在等待队列中，第 4 个客户端会得到错误信息。

4. connect 函数请求连接

connect 函数用于发送一个连接请求。如果函数执行成功，返回值为 0，否则为 SOCKET_ERROR。用户可以通过 WSAGetLastError 得到其错误描述。

语法为

int connect (SOCKET s, const struct sockaddr FAR *name, int namelen);

s：标识一个套接字。

name：套接字 s 想要连接的主机地址和端口号。

namelen：name 缓冲区的长度。

5. accept 函数接受连接

accpet 函数用于接受客户端的连接请求。返回值是一个新的套接字，它对应于已经接受的客户端连接，对于该客户端的所有后续操作，都应使用这个新的套接字。

语法为

    SOCKET accept（SOCKET s，struct sockaddr FAR ＊addr，
         int FAR ＊addrlen）；

s：一个套接字，它应处于监听状态。
addr：一个 sockaddr_in 结构指针，包含一组客户端的端口号、IP 地址等信息。
addrlen：用于接收参数 addr 的长度。

6. close socket 函数关闭连接

closesocket 函数用于关闭某个套接字。

语法为

       int closesocket（SOCKET s）；

s：标识一个套接字。如果参数 s 设置有 SO_DONTLINGER 选项，则调用该函数后会立即返回，但此时如果有数据尚未传送完毕，会继续传递数据，然后才关闭套接字。

6.5.2 send 与 recv 函数

1. send 函数

send 函数在已经建立连接的套接字上发送数据。
语法为
  int send（SOCKET s, const char FAR ＊buf, int len, int flags）；
send 函数参数说明如表 6-1 所示。

表 6-1　send 函数参数说明

参数名称	参数描述
s	标识一个套接字
buf	存放要发送数据的缓冲区
len	标识缓冲区长度
flags	标识函数的调用方式

2. recv 函数

recv 函数用于从连接的套接字中返回数据。
语法为

  int recv (SOCKET s, char FAR * buf, int len, int flags);

recv 函数参数的说明如表 6-2 所示。

表 6-2　recv 函数参数说明

参数名称	参数描述
s	标识一个套接字
buf	接收数据的缓冲区
len	buf 的长度
flags	表示函数的调用方式，可选值如下：MSG_PEEK_用来查看传来的数据，在序列前端的数据会被复制一份到返回缓冲区中，但是这个数据不会从序列中移走 MSG_OOB_用来处理 Out-Of-Band 数据

6.6　TCP 套接字编程

本次编程实验选择在 Windows 环境下进行。样例程序演示了 Windows 平台中，用 C 语言实现 TCP 套接字通信的流程，程序只是简单地建立连接后发送一个消息，并关闭连接。

1. 客户端源程序

```
#include< Winsock2.h>
#include< stdio.h>
int main()
{
    //第一步:加载 socket 库函数
    //*********************************************************
    WORD wVersionRequested;
    WSADATA wsaData;
    int err;
    wVersionRequested= MAKEWORD( 1, 1 );
    err= WSAStartup( wVersionRequested, &wsaData );
    if ( err! = 0 ){
        return 0;
    }
```

```
if ( LOBYTE( wsaData.wVersion )! = 1 || HIBYTE( wsaData.wVersion )! = 1 ){
        WSACleanup( );
        return 0;
}
// **********************************************************

//第一步,创建套接字
SOCKET sockClient= socket(AF_INET,SOCK_STREAM,0) ;
//定义套接字地址
SOCKADDR_IN addrSrv;
addrSrv.sin_addr.S_un.S_addr= inet_addr("127.0.0.1");   //获取服务器 IP 地址,
inet_addr()将 IP 地址转为点分十进制的格式。
addrSrv.sin_family= AF_INET;   //sin_family 表示地址族,对于 IP 地址,sin_family
成员将一直是 AF_INET
addrSrv.sin_port= htons(1234);
//连接服务器
if(connect(sockClient,(SOCKADDR *)&addrSrv,sizeof(SOCKADDR))! = 0)
{
    //MessageBox("连接失败");
    //return;
    printf("error");
    return 0;
}else
{
    printf("success");
}
    char recvBuf[100];
    recv(sockClient,recvBuf,100,0);
    printf("客户端接收到的数据:% s",recvBuf);
    send(sockClient,"client send ",strlen("client send ")+ 1,0);
    //关闭套接字
    closesocket(sockClient);
    //清除套接字资源
    WSACleanup();
    return 0;
}
```

2. 服务器端源程序

```
#include< Winsock2.h>
```

```c
#include<stdio.h>
int main()
{
    //第一步:加载socket库函数
    //************************************************************
    WORD wVersionRequested;
    WSADATA wsaData;
    int err;
    wVersionRequested= MAKEWORD( 1, 1 );
    err= WSAStartup( wVersionRequested, &wsaData );
    if ( err!=0 ){
        return 0;
    }
    if ( LOBYTE( wsaData.wVersion )!=1 ||
         HIBYTE( wsaData.wVersion )!=1 ){
        WSACleanup( );
        return 0;
    }
    //************************************************************
    //第二步创建套接字
    SOCKET sockSrv= socket(AF_INET,SOCK_STREAM,0);
    //第三步:绑定套接字
    //获取地址结构
    SOCKADDR_IN addrSrv;
    addrSrv.sin_addr.S_un.S_addr= htonl(INADDR_ANY);
    //将IP地址指定为INADDR_ANY,允许套接字向任何分配给本地机器的IP地址发送
    或接收数据
    //htonl()将主机的无符号长整形数转换成网络字节顺序。
    addrSrv.sin_family= AF_INET;
    //sin_family表示地址族,对于IP地址,sin_family成员将一直是AF_INET
    addrSrv.sin_port= htons(6000);
    //htons()将主机的无符号短整形数转换成网络字节顺序
    bind(sockSrv,(SOCKADDR *)&addrSrv,sizeof(SOCKADDR));
    //监听客户端
    listen(sockSrv,5);
    //定义从客户端接受的地址信息
    SOCKADDR_IN addrClient ;
    int len= sizeof(SOCKADDR);
    while(1)
```

```
{
    //不断等待客户端的请求的到来,并接受客户端的连接请求
    printf("等待客户连接\n");
    SOCKET sockConn= accept(sockSrv,(SOCKADDR *)&addrClient,&len);
    char sendBuf[100];
    sprintf(sendBuf,"welcome %s to snnu",inet_ntoa(addrClient.sin_addr));
    printf("发送数据\n");
    send(sockConn,sendBuf,strlen(sendBuf)+ 1,0);
    char recvBuf[100];
    printf("等待接收数据\n");
    recv(sockConn,recvBuf,100,0);
    printf("% s\n",recvBuf);
    closesocket(sockConn);
}
WSACleanup();
return 0;
}
```

6.7 UDP 套接字编程

样例程序实现的功能和 TCP 套接字例程相同。UDP 套接字也分为客户端和服务器两部分。

1. 客户端源程序

```
#include "Winsock2.h"
#pragma comment(lib,"ws2_32.lib")
#include "stdio.h"
int _tmain(int argc, _TCHAR *argv[])
{
    //第一步:加载 socket 库函数
    //***************************************************
    WORD wVersionRequested;
    WSADATA wsaData;
    int err;
    wVersionRequested= MAKEWORD( 1, 1 );
    err= WSAStartup( wVersionRequested, &wsaData );
    if ( err != 0 ){
        return 0;
```

```
        }
        if ( LOBYTE( wsaData.wVersion )! = 1 ||
            HIBYTE( wsaData.wVersion )! = 1 ){
                WSACleanup( );
                return 0;
        }
        //创建套接字
        SOCKET sockClient= socket(AF_INET,SOCK_DGRAM,0);
        SOCKADDR_IN sockSrv;
        sockSrv.sin_addr.S_un.S_addr= inet_addr("127.0.0.1");
        sockSrv.sin_family= AF_INET;
        sockSrv.sin_port= htons(6000);
        sendto(sockClient,"hello",strlen("hello")+ 1,0,(SOCKADDR *)&sockSrv,sizeof(SOCKADDR));
        closesocket(sockClient);
        WSACleanup();
        return 0;
    }
```

2. 服务器端源程序

```
    #include "Winsock2.h"
    #pragma comment(lib,"ws2_32.lib")
    #include "stdio.h"
    int _tmain(int argc, _TCHAR * argv[])
    {
        //第一步:加载 socket 库函数
        //************************************************************
        WORD wVersionRequested;
        WSADATA wsaData;
        int err;
        wVersionRequested= MAKEWORD( 1,1 );
        err= WSAStartup( wVersionRequested, &wsaData );
        if ( err ! = 0 ){
            return 0;
        }
        if ( LOBYTE( wsaData.wVersion )! = 1 ||
            HIBYTE( wsaData.wVersion )! = 1 ){
                WSACleanup( );
```

```
    return 0;
}
//创建套接字
SOCKET sockSrv= socket(AF_INET,SOCK_DGRAM,0);
SOCKADDR_IN addSrv;
addSrv.sin_addr.S_un.S_addr= inet_addr("127.0.0.1");
addSrv.sin_family= AF_INET;
addSrv.sin_port= htons(6000);
bind(sockSrv,(SOCKADDR *)&addSrv,sizeof(SOCKADDR));
SOCKADDR_IN addrClient;
int len= sizeof(SOCKADDR);
char recvBuf[100];
recvfrom(sockSrv,recvBuf,100,0,(SOCKADDR *)&addrClient,&len);
printf("%s\n",recvBuf);
closesocket(sockSrv);
WSACleanup();
return 0;
```

第 7 章　多线程 Web 服务器

7.1　实验目标

本实验将开发一个简单的 Web 多线程服务器，能处理简单的 Web 页面请求，接收 httpRequest 消息，并经过分析后构造响应消息，然后交付给浏览器。该服务器可以处理各种各样的对象，不仅包括 HTML 页面，而且也包括图像，还能连接子页面。

7.2　系统设计与组成

系统由两个类构成，分别实现了不同的功能。

WebHttpRequest：解析和处理来自客户端的传入请求，从服务器读取答复并进行处理。

WebServer：启动服务器并且新建线程处理请求。

首先开启服务器，并创建一个欢迎套接字监听指定端口，然后浏览器通过"浏览器 socket"发起到服务器指定端口的连接，此时服务器创建一个连接套接字与之通信并接受浏览器提交的请求，服务器找到待找页面后通过 socket 将结果返回给浏览器显示，如图 7-1 所示。

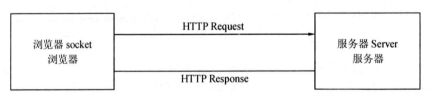

图 7-1　Web 服务器系统设计

7.3　重要类及方法

Public Socket（String host，int port）Throws UnknownHostException，IOException 创建一个流套接字并将其连接到指定主机上的指定端口号。

Public ServerSocket（　）Throws IOException 创建非绑定服务器套接字。

Public Socket accept（　）Throws IOException 侦听并接收到此套接字的连

接。此方法在连接传入之前一直阻塞。

Public StringTokenizer（String str）为指定字符串构造一个 String tokenizer。tokenizer 使用默认的分隔符集"\t\n\r\f"，即空白字符、制表符、换行符、回车符和换页符。分隔符字符本身不作为标记。

Public String nextToken（　）返回此 String tokenizer 的下一个标记。

Private void processRequest（　）throws Exception 接受从指定端口号的流，从中提取 URL 等关键信息，从本地服务器搜寻文件，并且构造新的流回复。

Private void sendBytes（FileInputStream fis，OutputStream os）throws Exception 将 os 流中的数据写到 fis 文件流中。

Private String contentType（String fileName）返回文件类型，有 html、htm、jpg 等类型。

7.4　开 发 环 境

此系统使用 JAVA 语言开发，运行在 Windows 平台下，使用前应安装并配置好相关开发环境。

7.5　运 行 结 果

在不接入 Internet 的情况下在浏览器地址栏输入 http：//127.0.0.1：6789/即可。

如输入 http：//127.0.0.1：6789/，结果如图 7-2 所示。

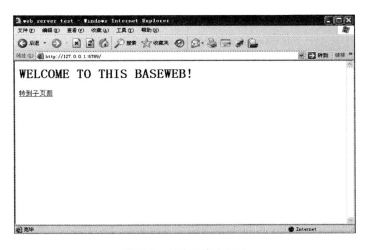

图 7-2　浏览器接入显示

单击"转到子页面",结果如图 7-3 所示。

图 7-3 跳转子页面

7.6 源 代 码

1. WebHttpRequest.java 文件

```
import java.io.BufferedReader;
import java.io.DataOutputStream;
import java.io.FileInputStream;
import java.io.FileNotFoundException;
import java.io.IOException;
import java.io.InputStream;
import java.io.InputStreamReader;
import java.io.OutputStream;
import java.net.Socket;
import java.util.StringTokenizer;
public class WebHttpRequest implements Runnable
{
    final static String CRLF= "\r\n";
    Socket socket;
    // Constructor
    public WebHttpRequest(Socket socket) throws Exception
    {
        this.socket= socket;
    }
```

```java
// Implement the run() method of the Runnable interface.
public void run()
{
    try
    {
        processRequest();
    }
    catch (Exception e)
    {
        System.out.println(e);
    }
}
private void processRequest() throws Exception
{
    // Get a reference to the socket's input and output streams.
    InputStream is= socket.getInputStream();
    DataOutputStream os= new DataOutputStream(socket.getOutputStream());
    // Set up input stream filters.
    BufferedReader br= new BufferedReader( newInputStreamReader( sock-
    et.getInputStream() ) ;
    // Get the request line of the HTTP request message.
    String requestLine= null;
    try
    {
        requestLine= br.readLine();
    }
    catch (IOException e)
    {
        System.out.println("Error reading request line: " + e);
    }
    // Get and display the header lines.
    String headerLine= null;
    // Display the request line.
    System.out.println();
    System.out.println(requestLine);
    while ((headerLine= br.readLine()).length() != 0)
    {
        System.out.println(headerLine);
    }
```

```java
// Extract the filename from the request line.
StringTokenizer tokens= new StringTokenizer(requestLine);
tokens.nextToken();// skip over the method, which should be "GET"
String fileName= tokens.nextToken();
String baseIndex= "Welcome.html";
//BASE WEB
// Prepend a "." so that file request is within the current directory.
if(fileName.equals("/"))
    fileName= "./" + baseIndex;
else
    fileName= "." + fileName;
// Open the requested file.
FileInputStream fis= null;
boolean fileExists= true;
try
{
    fis= new FileInputStream(fileName);
}
catch (FileNotFoundException e)
{
    fileExists= false;
    e.printStackTrace();
}
// Construct the response message.
String statusLine= null;
String contentTypeLine= null;
String entityBody= null;
String serverLine= "Server: a simple java httpServer";
String contentLengthLine= "error";
if(fileName.indexOf("..")!=-1)
    fileExists= false;
if ( fileExists )
{
    statusLine= "HTTP/1.1 200 OK" + CRLF;
    contentTypeLine= "Content- type: " + contentType(fileName) + CRLF;
    contentLengthLine = " Content - Length: " + (new Integer
    (fis.available())).toString() + CRLF;
}
else
```

```
        {
            statusLine= "HTTP/1.1 404 Not Found" + CRLF;
            contentTypeLine= "text/html" + CRLF;
            entityBody= "< HTML> " + "< HEAD> < TITLE> 404 Not Found< /TI-
            TLE> < /HEAD> " + "< BODY> 404 Not Found"+ "< br> usage:http://
            yourHostName:port/"+ "fileName.html< /BODY> < /HTML> " + CRLF;
        }
        // 发送到服务器信息
        os.write(statusLine.getBytes());
        os.write(serverLine.getBytes());
        os.write(contentTypeLine.getBytes());
        os.write(contentLengthLine.getBytes());
        os.write(CRLF.getBytes());
        // 发送信息内容
        if (fileExists)
        {
            sendBytes(fis, os);
            fis.close();
        }
        else
        {
            os.writeBytes( entityBody );
        }
        // Close streams and socket.
        os.close();
        br.close();
        socket.close();
    }
    private void sendBytes(FileInputStream fis, OutputStream os) throws Exception
    {
        // Construct a 1K buffer to hold bytes on their way to the socket.
        byte[] buffer= new byte[1024];
        int bytes= 0;
        // Copy requested file into the socket's output stream.
        try
        {
            while((bytes= fis.read(buffer))! = - 1 )
            {
                os.write(buffer, 0, bytes);
```

```java
                    }
                }
                catch (Exception e)
                {
                    e.printStackTrace();
                }
            }
            private String contentType(String fileName)
            {
                if(fileName.endsWith(".htm")||fileName.endsWith(".html"))
                {
                    return "text/html";
                }
                if(fileName.endsWith(".jpg")||fileName.endsWith(".JPG")||
                fileName.endsWith(".Jpg"))
                {
                    return "image/jpg";
                }
                if(fileName.endsWith(".gif")||fileName.endsWith(".GIF")||
                fileName.endsWith(".Gif"))
                {
                    return "image/gif";
                }
                return "application/octet-stream";
            }
        }
```

2. webServer.java 文件

```java
    import java.io.*;
    import java.net.*;
    import java.util.*;
    public class webServer
    {
        public static void main(String argv[]) //throws Exception
        {
            final int PORT= 6789;
            // Establish the listen socket.
            ServerSocket server= null;
```

```java
        // Process HTTP service requests in an infinite loop.
        try
        {
            server = new ServerSocket( PORT );
            System.out.println( " Http Server is running on port " + server.getLocalPort()
                    + ", address: " + server.getInetAddress() );
            while (true)
            {
                // Listen for a TCP connection request.
                Socket socket = server.accept();
                // Construct an object to process the HTTP request message.
                WebHttpRequest request = new WebHttpRequest( socket );
                // Create a new thread to process the request.
                Thread thread = new Thread(request);
                thread.start();
            }
        }
        catch (IOException e)
        {
            e.printStackTrace();
        } catch (Exception e) {
            // TODO 自动生成 catch 块
            e.printStackTrace();
        }
        finally
        {
            try
            {
                server.close();
            }
            catch (IOException e)
            {
                e.printStackTrace();
            }
        }
    }
}
```

第 8 章　邮件客户端

8.1　实验目标

设计一个简单的邮件客户端 MUA，为发送者提供了一个图形界面，其中具有用于本地邮件服务器的字段，即 SMTP 服务器主机名、发送者电子邮件地址、接收者电子邮件地址、报文主题及报文本身。

设计思路是在邮件客户机和本地邮件服务器之间创建一个 TCP 连接，向本地邮件服务器发送 SMTP 命令并从本地邮件服务器接收和处理 SMTP 命令。

邮件客户端前端界面如图 8-1 所示。

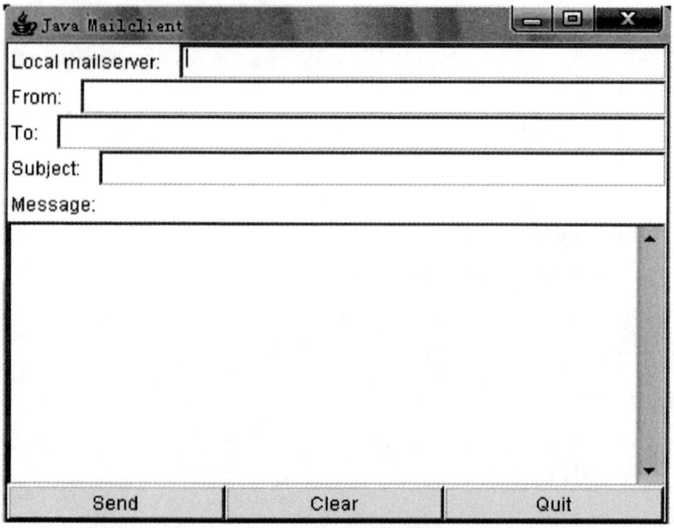

图 8-1　邮件客户端前端界面

每次至多给一个接收者发送电子邮件。并且，该用户代理需要 SMTP 服务器规范名字。

系统由四个类构成，分别实现了不同的功能。

MailClient：提供主界面。

Message：封装要发送的信息。

Envelope：信息封装（发送地址、接收地址）。

SMTPConnection：建立连接和服务器通信。

8.2 系统设计与组成

程序中，先通过主界面接收用户的输入，再将邮件基本信息封装在 Message 中，然后加上地址信息等邮件首部封装进 Envelope，最后利用 SMTPConnection 和 SMTP 服务器建立连接，根据 SMTP 协议发送邮件，在发送的过程中如果有异常直接抛出，发送失败，否则发送成功。

8.3 重要类及方法

Message 类主要有四个属性，分别如下。
public String Headers：邮件首部。
public String Body：邮件主体。
private String From：发件地址。
private String To：收件地址。
Message 类中还有一些处理函数，isValid（ ）是用来判断邮件地址的格式是否正确的，在这类中，对邮件首部和内容做了一些格式上的处理，以达到 RFC 中的要求。

Envelope 是将 Message 进一步封装的类，主要属性如下。
public String Sender：发件地址。
public String Recipient：收件地址。
public String DestHost：SMTP 服务器主机名。
public InetAddress DestAddr：SMTP 服务器 IP 地址。
public Message Message：Message 类对象。
Envelope 中还有用来帮助调试的函数：
　　　　/ * For printing the envelope. Only for debug. * /
　　　　public String toString（ ）

SMTPConnection 是其中最关键的类。sendCommand 是其中最重要的函数，函数有两个参数，private void sendCommand（String command, int rc）。第一个是要发送的命令（如 MAIL FROM:someboby@snnu.edu.cn），第二个是正确情况下的状态码。函数会先发送命令，再读取返回信息，判断返回的状态码和正确情况下的是否一样，不一样就抛出异常。

8.4 开 发 环 境

此系统使用 JAVA 语言开发，运行在 Windows 平台下，使用前应安装并配置好相关开发环境。

8.5 运行结果

输入如图 8-2 所示的内容。

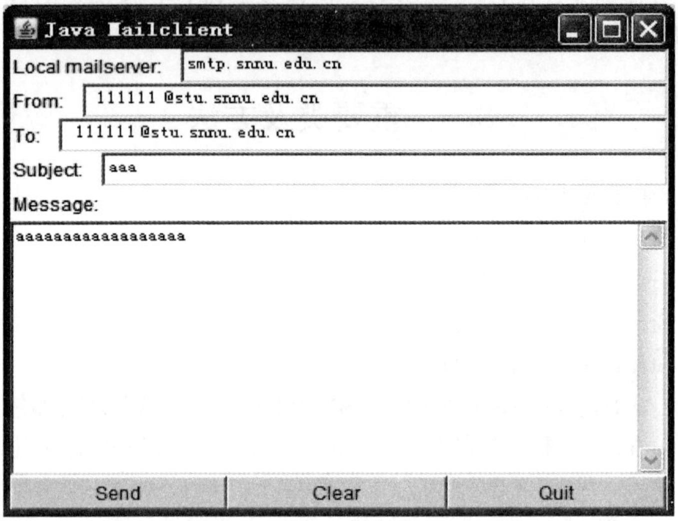

图 8-2 邮箱界面

邮件接收成功验证如图 8-3 所示。

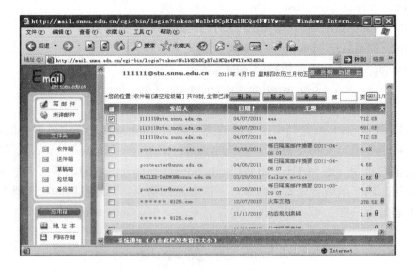

图 8-3 邮箱验证界面

打开该邮件如图 8-4 所示。

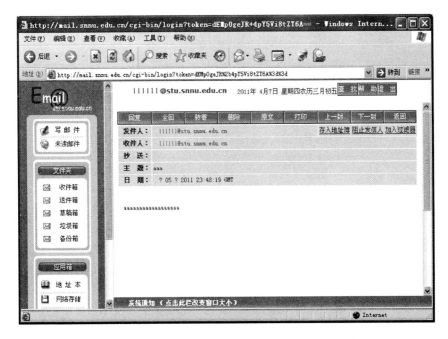

图 8-4 打开邮件界面

8.6 源 代 码

1. MailClient.java 文件

```
import java.io.*;
import java.net.*;
import java.awt.*;
import java.awt.event.*;
public class MailClient extends Frame
{
    /* The stuff for the GUI.*/
    private Button btSend= new Button("Send");
    private Button btClear= new Button("Clear");
    private Button btQuit= new Button("Quit");
    private Label serverLabel= new Label("Local mailserver:");
    private TextField serverField= new TextField("", 40);
    private Label fromLabel= new Label("From:");
```

```java
private TextField fromField= new TextField("", 40);
private Label toLabel= new Label("To:");
private TextField toField= new TextField("", 40);
private Label subjectLabel= new Label("Subject:");
private TextField subjectField= new TextField("", 40);
private Label messageLabel= new Label("Message:");
private TextArea messageText= new TextArea(10, 40);
/**
 * Create a new MailClient window with fields for entering all
 * the relevant information (From, To, Subject, and message).
 */
public MailClient()
{
    super("Java Mailclient");
    /* Create panels for holding the fields. To make it look nice,
       create an extra panel for holding all the child panels.*/
    Panel serverPanel= new Panel(new BorderLayout());
    Panel fromPanel= new Panel(new BorderLayout());
    Panel toPanel= new Panel(new BorderLayout());
    Panel subjectPanel= new Panel(new BorderLayout());
    Panel messagePanel= new Panel(new BorderLayout());
    serverPanel.add(serverLabel, BorderLayout.WEST);
    serverPanel.add(serverField, BorderLayout.CENTER);
    fromPanel.add(fromLabel, BorderLayout.WEST);
    fromPanel.add(fromField, BorderLayout.CENTER);
    toPanel.add(toLabel, BorderLayout.WEST);
    toPanel.add(toField, BorderLayout.CENTER);
    subjectPanel.add(subjectLabel, BorderLayout.WEST);
    subjectPanel.add(subjectField, BorderLayout.CENTER);
    messagePanel.add(messageLabel, BorderLayout.NORTH);
    messagePanel.add(messageText, BorderLayout.CENTER);
    Panel fieldPanel= new Panel(new GridLayout(0, 1));
    fieldPanel.add(serverPanel);
    fieldPanel.add(fromPanel);
    fieldPanel.add(toPanel);
    fieldPanel.add(subjectPanel);
    /* Create a panel for the buttons and add listeners to the
       buttons.*/
    Panel buttonPanel= new Panel(new GridLayout(1, 0));
```

```java
        btSend.addActionListener(new SendListener());
        btClear.addActionListener(new ClearListener());
        btQuit.addActionListener(new QuitListener());
        buttonPanel.add(btSend);
        buttonPanel.add(btClear);
        buttonPanel.add(btQuit);

        /* Add, pack, and show.*/
        add(fieldPanel, BorderLayout.NORTH);
        add(messagePanel, BorderLayout.CENTER);
        add(buttonPanel, BorderLayout.SOUTH);
        pack();
        show();
    }
    static public void main(String argv[])
    {
        new MailClient();
    }
/* Handler for the Send-button.*/
class SendListener implements ActionListener
{
    public void actionPerformed(ActionEvent event)
    {
      System.out.println("Sending mail");
      /* Check that we have the local mailserver */
        if((serverField.getText()).equals(""))
        {
            System.out.println("Need name of local mailserver!");
            return;
        }
        /* Check that we have the sender and recipient.*/
        if((fromField.getText()).equals(""))
        {
            System.out.println("Need sender!");
            return;
        }
    if((toField.getText()).equals(""))
        {
            System.out.println("Need recipient!");
```

```
        return;
    }
    /* Create the message */
    Message mailMessage= new Message(fromField.getText(),
                    toField.getText(),
                    subjectField.getText(),
                    messageText.getText());
    /* Check that the message is valid, i.e., sender and
        recipient addresses look ok.*/
    if(! mailMessage.isValid())
    {
        return;
    }
    /* Create the envelope, open the connection and try to send
        the message.*/
    Envelope envelope;
    try {
            envelope= new Envelope(mailMessage,
                            serverField.getText());
    }
    catch (UnknownHostException e)
    {
        /* If there is an error, do not go further */
        return;
    }
    try {
        SMTPConnection connection= new SMTPConnection(envelope);
            connection.send(envelope);
            connection.close();
    }
    catch (IOException error)
        {
            System.out.println("Sending failed: " + error);
            return;
        }
    System.out.println("Mail sent succesfully!");
  }
}
    /* Clear the fields on the GUI.*/
```

```java
class ClearListener implements ActionListener
{
    public void actionPerformed(ActionEvente)
    {
        System.out.println("Clearing fields");
        fromField.setText("");
        toField.setText("");
        subjectField.setText("");
        messageText.setText("");
    }
}
/* Quit. */
class QuitListener implements ActionListener
{
    public void actionPerformed(ActionEvente)
    {
        System.exit(0);
    }
}
}
```

2. Message.java 文件

```java
import java.util.*;
import java.text.*;
public class Message
{
    /* The headers and the body of the message.*/
    public String Headers;
    public String Body;
    /* Sender and recipient. With these, we don't need to extract them
        from the headers.*/
    private String From;
    private String To;
    /* To make it look nicer */
    private static final String CRLF= "\r\n";
    /* Create the message object by inserting the required headers from
        RFC 822 (From, To, Date).*/
    public Message(String from, String to, String subject, String text)
```

```java
        {
            /* Remove white space */
            From= from.trim();
            To= to.trim();
            Headers= "From: " + From + CRLF;
            Headers += "To: " + To + CRLF;
            Headers += "Subject: " + subject.trim() + CRLF;
            /* A close approximation of the required format. Unfortunately
               only GMT.*/
            SimpleDateFormat format=
                new SimpleDateFormat("EEE, dd MMM yyyy HH:mm:ss 'GMT'");
            String dateString= format.format(new Date());
            Headers += "Date: " + dateString + CRLF;
            Body= text;
        }
    /* Two functions to access the sender and recipient.*/
    public String getFrom()
        {
            return From;
        }
    public String getTo()
        {
            return To;
        }
    /* Check whether the message is valid. In other words, check that
       both sender and recipient contain only one @-sign.*/
    public boolean isValid()
        {
            int fromat= From.indexOf('@');
            int toat= To.indexOf('@');
            if(fromat < 1 || (From.length() - fromat) <= 1)
              {
                System.out.println("Sender address is invalid");
                return false;
              }
            if(toat < 1 || (To.length() - toat) <= 1)
              {
                System.out.println("Recipient address is invalid");
                return false;
```

```
                }
                if(fromat ! = From. lastIndexOf('@'))
                  {
                        System. out. println("Sender address is invalid");
                        return false;
                  }
                if(toat ! = To. lastIndexOf('@'))
                  {
                        System. out. println("Recipient address is invalid");
                        return false;
                  }
            return true;
        }
    /* For printing the message.*/
    public String toString()
        {
            String res;
            res= Headers +  CRLF;
            res + = Body;
            return res;
        }
}
```

3. Envelope.java 文件

```
    import java. io.*;
    import java. net.*;
    import java. util.*;
    public class Envelope
    {
        public String Sender;
        public String Recipient;
        public String DestHost;
        public InetAddress DestAddr;
        public Message Message;
        public Envelope(Message message, String localServer) throws UnknownHostException
            {Sender= message. getFrom();
             Recipient= message. getTo();
             Message= escapeMessage(message);
```

```
            DestHost= localServer;
        try { DestAddr= InetAddress.getByName(DestHost);
            }
        catch (UnknownHostException e)
            {System.out.println("Unknown host: " + DestHost);
            System.out.println(e);
            throw e;
            }
            return;
        }
    private Message escapeMessage(Message message)
        {
            String escapedBody= "";
            String token;
            StringTokenizer parser= new StringTokenizer(message.Body, "\n", true);
          while(parser.hasMoreTokens())
            {
            token= parser.nextToken();
            if(token.startsWith("."))
                {
                token= "." + token;
                }
                escapedBody + = token;
            }
            message.Body= escapedBody;
            return message;
         }
         /* For printing the envelope. Only for debug.*/
        public String toString()
        {String res= "Sender: " + Sender + '\n';
            res + = "Recipient: " + Recipient + '\n';
            res + = "MX-host: " + DestHost + ", address: " + DestAddr + '\n';
            res + = "Message:" + '\n';
            res + = Message.toString();
            return res;
          }
     }
```

4. SMTPConnection.java 文件

```
    import java.net.*;
```

```java
import java.io.*;
import java.util.*;
public class SMTPConnection
    {
        private Socket connection;
        private BufferedReader fromServer;
        private DataOutputStream toServer;
        private static final int SMTP_PORT= 25;
        private static final String CRLF= "\r\n";
        private boolean isConnected= false;
        public SMTPConnection(Envelope envelope) throws IOException
            {
                Envelope e= envelope;
                connection= new Socket(e.DestHost,SMTP_PORT);
                fromServer= new BufferedReader(new InputStreamReader(connec-
                tion.getInputStream()));
                toServer= new DataOutputStream( connection.getOutputStream());
                String b;
                b= fromServer.readLine();
                int a= parseReply(b);
                if (a! = 220)
                {
                    throw (new IOException());
                }
                InetAddress myip= InetAddress.getByName("");
                String localhost = myip.getHostAddress();
                sendCommand( "HELO"+ ' '+ localhost + CRLF,250 );
                isConnected= true;
            }
        public void send(Envelope envelope) throws IOException
            {
                Envelope e= envelope;
                sendCommand("MAIL FROM: < " + e.Sender + '>'+ CRLF,250);
                sendCommand("RCPT TO: < " + e.Recipient + '>'+ CRLF,250);
                sendCommand("DATA" + CRLF,354);
                //传送message,用套接字
                toServer.writeBytes(e.Message.Headers + CRLF);
                toServer.writeBytes(e.Message.Body + CRLF+ '.'+ CRLF);
                fromServer.readLine();
```

```java
        }
/* Close the connection. First, terminate on SMTP level, then
   close the socket.*/
        public void close()
          {
              isConnected= false;
              try {
                  sendCommand("QUIT" + CRLF,221);
                  connection.close();
               }
              catch (IOException e)
                {
                    System.out.println("Unable to close connection: " + e);
                    isConnected= true;
                }
          }
/* Send an SMTP command to the server. Check that the reply code is
   what is is supposed to be according to RFC 821.*/
        private void sendCommand(String command, int rc) throws IOException
             {
              String b;
              toServer.writeBytes(command);
              b= fromServer.readLine();
              int a= parseReply(b);
              if(a! = rc)
                {
                    throw (new IOException());
                }
             }
        private int parseReply(String reply)
            {
              String a= reply;
              String b;
              b= a.substring(0,3);
              int i= Integer.parseInt(b);
              return i;
            }
        protected void finalize() throws Throwable
            {
```

```
            if(isConnected)
        {
            close();
        }
    super.finalize();
    }
}
```

第9章 邮件用户代理：控制台版本

9.1 实验目标

本系统分为两部分，第一部分用 Telnet 通过一个 SMTP 邮件服务器实现邮件发送；第二部分用 JAVA 代码实现该功能。

9.2 系统设计与组成

1. 第一部分：用 Telnet 发邮件

实现步骤如下。

(1) 首先建立 TCP 连接，命令为 telnet smtp.snnu.edu.cn 25，如图 9-1 所示。

图 9-1　建立连接

(2) 其次输入相应的邮件传输命令，具体如图 9-2 所示。

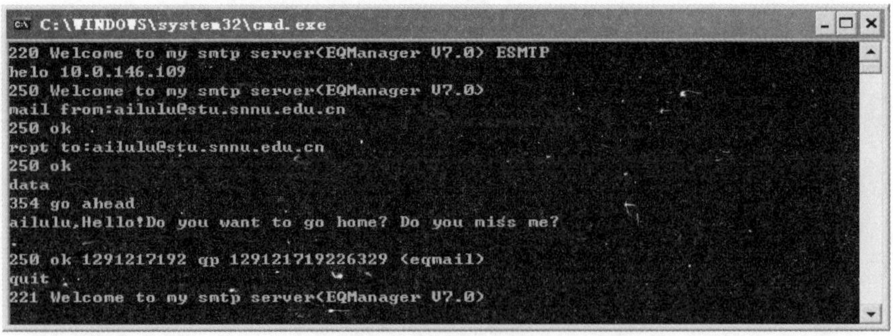

图 9-2　邮件传输命令

（3）最后检查邮箱发现信件投递成功。

2. 第二部分：用 JAVA 程序发送邮件

JAVA 提供了一个称为 JavaMail 的 API 和邮件系统。本实验不用此 API，因为它隐藏了 SMTP 和 socket 编程的细节。

本系统把所有代码封装进了一个 EmailSender 类中，并且将所要发送的信息也直接写在代码中。实现流程如下。

（1）建立 TCP 连接。
（2）建立一个 BufferedReader 实现一次读一行。
（3）从服务器读出成功连接的回复代码。
（4）发送 HELO 命令并获得服务器的回复。
（5）发送 MAIL FROM、RCPT TO、DATA、QUIT 等命令并获取服务器相应回复代码。
（6）邮件成功传输。

9.3 重要的类及实现

DateFormat：使用 getDateInstance 来获取该国家/地区的标准日期格式。

public Socket（String host，int port）Throws UnknownHostException, IOException 创建一个流套接字并将其连接到指定主机上的指定端口号。

public class BufferedReader extends Reader 从字符输入流中读取文本，缓冲各个字符，从而实现字符、数组和行的高效读取。

9.4 开发环境

此系统使用 JAVA 语言开发，运行在 Windows 平台下，使用前应安装并配置好相关开发环境。

9.5 运行结果

检查邮箱时，信件投递成功，如图 9-3 显示。

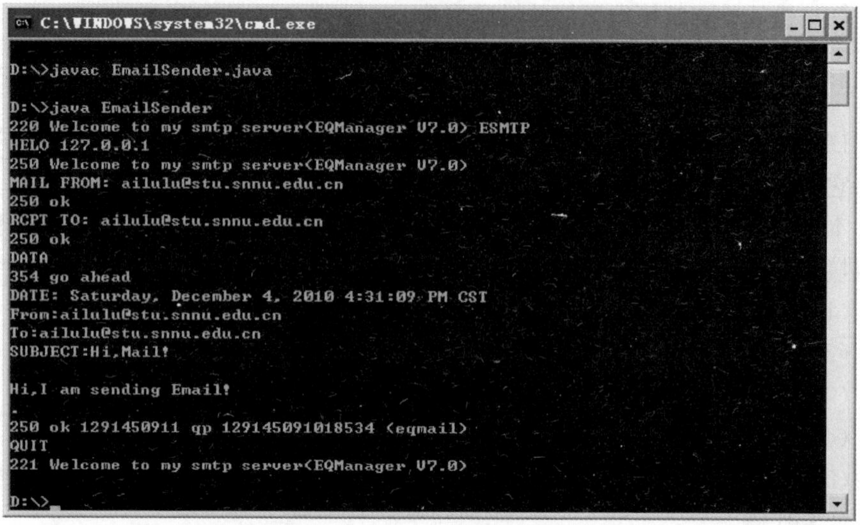

```
C:\WINDOWS\system32\cmd.exe

D:\>javac EmailSender.java

D:\>java EmailSender
220 Welcome to my smtp server(EQManager V7.0) ESMTP
HELO 127.0.0.1
250 Welcome to my smtp server(EQManager V7.0)
MAIL FROM: ailulu@stu.snnu.edu.cn
250 ok
RCPT TO: ailulu@stu.snnu.edu.cn
250 ok
DATA
354 go ahead
DATE: Saturday, December 4, 2010 4:31:09 PM CST
From:ailulu@stu.snnu.edu.cn
To:ailulu@stu.snnu.edu.cn
SUBJECT:Hi,Mail!

Hi,I am sending Email!

250 ok 1291450911 qp 129145091018534 (eqmail)
QUIT
221 Welcome to my smtp server(EQManager V7.0)

D:\>
```

图 9-3　邮箱信件投递成功

9.6　源　代　码

EmailSender.java 文件

```
import java.net.*;
import java.io.*;
import java.util.*;
public class SMTPConnection
    {
    private Socket connection;
    private BufferedReader fromServer;
    private DataOutputStream toServer;
    private static final int SMTP_PORT= 25;
    private static final String CRLF= "\r\n";
    private boolean isConnected= false;
    public SMTPConnection(Envelope envelope) throws IOException
        {
        Envelope e= envelope;
        connection= new Socket(e.DestHost,SMTP_PORT);
        fromServer= new BufferedReader(new
        InputStreamReader(connection.getInputStream()) );
        toServer= new DataOutputStream( connection.getOutputStream();
```

```java
            String b;
            b= fromServer.readLine();
            int a= parseReply(b);
            if (a! = 220)
            {
               throw (new IOException());
            }
            InetAddress myip= InetAddress.getByName("");
            String localhost = myip.getHostAddress();
            sendCommand( "HELO"+ ' '+ localhost + CRLF,250 );
            isConnected= true;
      }
   public void send(Envelope envelope) throws IOException
         {
            Envelope e= envelope;
            sendCommand("MAIL FROM: < " + e.Sender + '>'+ CRLF,250);
            sendCommand("RCPT TO: < " + e.Recipient + '>'+ CRLF,250);
            sendCommand("DATA" + CRLF,354);
            //传送 message,用套接字
            toServer.writeBytes(e.Message.Headers + CRLF);
            toServer.writeBytes(e.Message.Body + CRLF+ '.'+ CRLF);
            fromServer.readLine();
         }
/* Close the connection. First, terminate on SMTP level, then
   close the socket.*/
public void close()
   {
         isConnected= false;
         try {
                  sendCommand("QUIT" + CRLF,221);
                  connection.close();
         }
      catch (IOException e)
         {
                  System.out.println("Unable to close connection: " + e);
                  isConnected= true;
         }
   }
   /* Send an SMTP command to the server. Check that the reply code is
```

what is is supposed to be according to RFC 821.*/
```
private void sendCommand(String command, int rc) throws IOException
    {
        String b;
        toServer.writeBytes(command);
        b= fromServer.readLine();
        int a= parseReply(b);
        if(a! = rc)
        {throw (new IOException());
        }
    }
private int parseReply(String reply)
    {
        String a= reply;
        String b;
        b= a.substring(0,3);
        int i= Integer.parseInt(b);
        return i;
    }
protected void finalize() throws Throwable
    {
        if(isConnected)
        {   close();
        }
        super.finalize();
    }
}
```

第 10 章　用 UDP 实现 ping 功能

10.1　实验目标

编写互联网 ping 服务器，并实现相应的客户端。需要完成的任务包括：
（1）实现客户端，使它发送 10 个 ping 请求到服务器，约 1s 间隔；
（2）每个消息包含数据的有效载荷，包括关键字 ping、序列号和时间戳；
（3）在发送每个数据包后，客户端最多等待 1s 收到答复。如果 1s 流逝，没有从服务器收到答复，然后在客户端假设其数据包或服务器的应答包已经在网络中丢失。

10.2　系统设计与组成

系统主要设计两个类，分别实现了不同的功能。
PingServer：解析和处理来自客户端的传入请求。
PingClient：发送 ping 探测消息并且处理服务器的应答。
功能上，系统主要分为两部分，一是服务器端，另一个是客户机端。这里主要是用 UDP 套接字连接的方法来实现一个简单的模拟 ping 程序。所使用的语言是 JAVA，模拟了丢包和网络延时，并实时显示 ping 的结果，实现了 ping 命令的要求。
在服务器端需要的参数为端口号，而在客户机端需要的参数有 ping 的 IP 地址、端口号及 ping 的次数。若只输入了 IP 地址，则默认 ping 10 次，若输入"-t"，则一直 ping 下去。

10.3　重要的类及实现

public DatagramSocket（int port）throws SocketException 用于创建数据报套接字并将其绑定到本地主机上的指定端口。套接字将被绑定到通配符地址，IP 地址由内核来选择。
public DatagramPacket（byte [] buf, int length, InetAddress address, int port）用于构造数据报包，用来将长度为 length 的包发送到指定主机上的指定端

口号。length 参数必须小于等于 buf.length。

public class BufferedReader extends Reader 从字符输入流中读取文本，缓冲各个字符，从而实现字符、数组和行的高效读取。

public ByteArrayInputStream（byte[] buf）创建一个 ByteArrayInputStream，使用 buf 作为其缓冲区数组。该缓冲区数组不是复制得到的。pos 的初始值是 0，count 的初始值是 buf 的长度。

10.4 开发环境

此系统使用 JAVA 语言开发，运行在 Windows 平台下，使用前应安装并配置好相关开发环境。

10.5 运行结果

服务器端运行情况如图 10-1 所示。

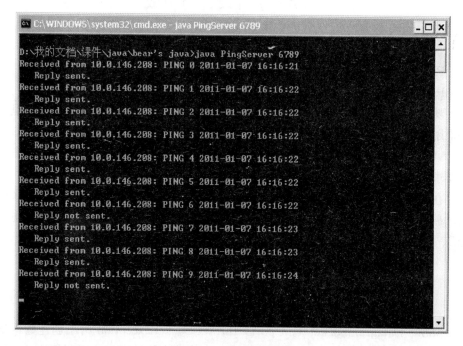

图 10-1 服务器端显示

客户机端运行情况如图 10-2 所示。

第 10 章　用 UDP 实现 ping 功能

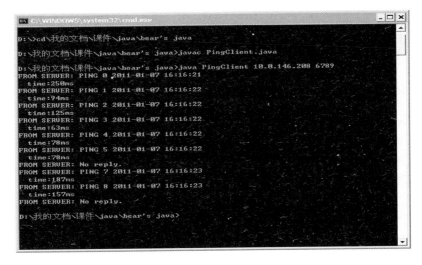

图 10-2　客户机端显示

可以进行特定次数的 ping。

服务器端如图 10-3 所示。

图 10-3　ping 服务器端显示

客户机端如图 10-4 所示。

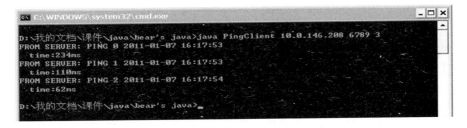

图 10-4　ping 客户机端显示

10.6 源代码

1. PingServer.java 文件

```java
import java.io.*;
import java.net.*;
import java.util.*;
/*
 * Server to process ping requests over UDP.
 */
public class PingServer
{
    private static final double LOSS_RATE= 0.3;
    private static final int AVERAGE_DELAY= 100;// milliseconds
    public static void main(String[] args) throws Exception
    {
        // Get command line argument.
        if (args.length ! = 1) {
            System.out.println("Required arguments: port");
            return;
        }
        int port= Integer.parseInt(args[0]);
        // Create random number generator for use in simulating
        // packet loss and network delay.
        Random random= new Random();
        // Create a datagram socket for receiving and sending UDP packets
        // through the port specified on the command line.
        DatagramSocket socket= new DatagramSocket(port);//创建套接字
        // Processing loop.
        while (true) {
            // Create a datagram packet to hold incomming UDP packet.
            DatagramPacket request= new DatagramPacket(new byte[1024], 1024);
            // Block until the host receives a UDP packet.
            socket.receive(request);
            // Print the recieved data.
            printData(request);
            // Decide whether to reply, or simulate packet loss.
```

```java
            if (random.nextDouble() < LOSS_RATE) {
                System.out.println("    Reply not sent.");
                continue;
            }
            // Simulate network delay.
            Thread.sleep((int) (random.nextDouble() * 2 * AVERAGE_DELAY));
            // Send reply.
            InetAddress clientHost = request.getAddress();//提取地址
            int clientPort = request.getPort();//提取端口号
            byte[] buf = request.getData();//提取数据
            DatagramPacket reply = new DatagramPacket(buf, buf.length, clientHost, client-
Port);//从 Buf 数组中,取出 Length 长的数据创建数据包对象,
            //目标是 clientHost 地址,clientPort 端口,通常用来发送数据给客户端。
            socket.send(reply);
            System.out.println("    Reply sent.");
        }
    }
    /*
     * Print ping data to the standard output stream.
     */
    private static void printData(DatagramPacket request) throws Exception
    {
        // Obtain references to the packet's array of bytes.
        byte[] buf = request.getData();
        // Wrap the bytes in a byte array input stream,
        // so that you can read the data as a stream of bytes.
        ByteArrayInputStream bais = new ByteArrayInputStream(buf);
        // Wrap the byte array output stream in an input stream reader,
        // so you can read the data as a stream of characters.
        InputStreamReader isr = new InputStreamReader(bais);
        // Wrap the input stream reader in a buffered reader,
        // so you can read the character data a line at a time.
        // (A line is a sequence of chars terminated by any combination of \r and \n.)
        BufferedReader br = new BufferedReader(isr);
        // The message data is contained in a single line, so read this line.
        String line = br.readLine();
        // Print host address and data received from it.
        System.out.println(
            "Received from " +
```

```
            request.getAddress().getHostAddress() + ": " + new String(line) );
        }
    }
```

2. PingClient.java 文件源代码

```java
import java.io.*;
import java.net.DatagramPacket;
import java.net.DatagramSocket;
import java.net.InetAddress;
import java.text.SimpleDateFormat;
import java.util.Date;
import java.util.Random;

public class PingClient{
    private static final double LOSS_RATE= 0.3;
    private static final int AVERAGE_DELAY= 100; //milliseconds

    public static void main(String []args) throws Exception{
        // Get command line argument.
        if ( args.length == 0 ){
            System.out.println("Required arguments: host port");
            return;
        }
        if ( args.length == 1 ){
            System.out.println("Required arguments: port");
            return;
        }

        String host= args[0].toString();
        int port= Integer.parseInt(args[1]);

        // try to link the Server
        DatagramSocket clientSocket= new DatagramSocket();
        // wait for 1 second
        clientSocket.setSoTimeout(1000);

        InetAddress IPAddress= InetAddress.getByName(host);
```

```java
for( int i=0; i<10; i++){
    // send the messages
    byte[] sendData= new byte[1024];
    byte[] receiveData= new byte[1024];
    Date currentTime= new Date();
    SimpleDateFormat formatter= new SimpleDateFormat("yyyy-MM-dd HH:mm:ss");
    String timeStamp= formatter.format(currentTime);
    String pingMessage= "PING" + i + " " + timeStamp + " " + "\r\n";
    sendData= pingMessage.getBytes();
    DatagramPacket sendPacket=new DatagramPacket
    (sendData,sendData.length,IPAddress,port );
    try{
        clientSocket.send(sendPacket);
        DatagramPacket receivePacket= new DatagramPacket
    (receiveData,receiveData.length );
        clientSocket.receive(receivePacket);
        String reply= new String(receivePacket.getData());
        System.out.println("FROM SERVER:" + reply);
    }catch( java.net.SocketTimeoutException ex){
        String reply= "No reply";
        System.out.println("FROM SERVER:" + reply);
    }
}
    // close the connection with Server
    clientSocket.close();
    }
}
```

第 11 章 Web 代理服务器

11.1 实验目标

本实验将开发一个简单的代理服务器，只能处理简单的 GET 请求，它能够接收 HTTP Request 消息，经过分析后将其转发，然后接收回复，最后交付给浏览器显示。该代理服务器应该可以处理各种各样的对象，不仅包括 HTML 网页，而且也包括图像。

11.2 系统设计与组成

系统由三个类构成，分别实现了不同的功能。
Http Request：解析和处理来自客户端的传入请求。
Http Response：从服务器读取答复并进行处理。
Proxy Cache：启动服务器并且处理提交的请求。
由于此系统涉及网络通信，所以选择 socket 作为开发系统的核心技术，socket 通信（发送 HTTP Request 和 Response 消息）如图 11-1 所示。

图 11-1 网络通信模型

首先开启代理服务器，并创建一个欢迎套接字监听指定端口，然后浏览器通过"浏览器 socket"发起到代理服务器指定端口的连接，此时代理服务器创建一个连接套接字"Client"与之通信并接受浏览器提交的 HTTP Request 消息，代理服务器解析 HTTP Request 消息并从请求行中得到要访问的主机地址，然后代理服务器通过"socket Server"发起到服务器的连接，服务器端监听到此连接后创建一个连接套接字"Serversocket"与之通信。

11.3 重要类及方法

public Socket (String host, int port) throws UnknownHostException, IOException 创建一个流套接字并将其连接到指定主机上的指定端口号。

public ServerSocket ( ) throws IOException 创建非绑定服务器套接字。

public Socket accept ( ) throws IOException 侦听并接受到此套接字的连接。在连接传入之前一直阻塞。

public static String toString (byte [] a) 返回指定数组内容的字符串表示形式。

public DataInputStream (InputStream in) 使用指定的底层 InputStream 创建一个 DataInputStream。

public final String readLine ( ) throws IOException 从包含的输入流中读取此操作需要的字节。

public final int read (byte [] b, int off, int len) throws IOException 从包含的输入流中将最多 len 个字节读入一个 byte 数组中。尽量读取 len 个字节,但读取的字节数可能少于 len 个,也可能为零。以整数形式返回实际读取的字节数。

public DataOutputStream (OutputStream out) 创建一个新的数据输出流,将数据写入指定基础输出流。计数器 written 被设置为零。

public final void writeBytes (String s) throws IOException 将字符串按字节顺序写出到基础输出流中。

11.4 开发环境

此系统使用 JAVA 语言开发,运行在 Windows 平台下,使用前应安装并配置好相关开发环境。此外运行该代理服务器需要设置客户端浏览器的代理服务器配置,方法为:选择菜单命令"工具→Internet 选项→连接→局域网设置→代理服务器",填入 '127.0.0.1,6666'(端口号可以任意设置,但要与命令行下输入的代理服务器端口一致)。

11.5 运 行 结 果

http://gaia.cs.umass.edu/ethereal-labs/HTTP-ethereal-file1.html 的运行结果如图 11-2 和图 11-3 所示。

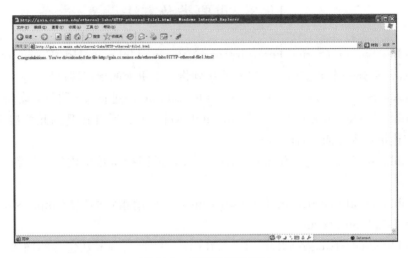

图 11-2　HTTP 网页显示

图 11-3　命令窗口显示

11.6 源 代 码

1. HttpRequest.java 文件

```java
/**
 * HttpRequest -HTTP request container and parser
 */
import java.io.*;
import java.net.*;
import java.util.*;
public class HttpRequest
{
    /** Help variables */
    final static String CRLF= "\r\n";
    final static int HTTP_PORT= 80;
    /** Store the request parameters */
    String method;
    String URI;
    String version;
    String headers= "";
    /** Server and port */
    private String host;
    private int port;
    /** Create HttpRequest by reading it from the client socket */
    public HttpRequest(BufferedReader from)
    {
        String firstLine= "";
        try
        {
            firstLine= from.readLine();
        }
        catch (IOException e)
        {
            System.out.println("Error reading request line: " + e);
        }
        String[] tmp= firstLine.split(" ");
        method = tmp[0];
```

```
            URI= tmp[1];
            version= tmp[2];
            System.out.println("URI is: " + URI);
            if (! method.equals("GET"))
            {
               System.out.println("Error: Method not GET");
            }
            try
            {
               String line= from.readLine();
               while (line.length() ! = 0)
               {
                    headers + = line + CRLF;
                    /* We need to find host header to know which server to
                     * contact in case the request URI is not complete.*/
                    if (line.startsWith("Host:"))
                    {
                       tmp= line.split(" ");
                       if (tmp[1].indexOf(':') > 0)
                       {
                           String[] tmp2= tmp[1].split(":");
                           host= tmp2[0];
                           port= Integer.parseInt(tmp2[1]);
                       }
                       else
                       {
                           host= tmp[1];
                           port= HTTP_PORT;
                       }
                    }
                    line= from.readLine();
               }
            }
            catch (IOException e)
            {
              System.out.println("Error reading from socket: " + e);
              return;
            }
            System.out.println("Host to contact is: " + host + " at port " + port);
```

```java
        }
        /** Return host for which this request is intended */
        public String getHost()
        {
          return host;
        }
        /** Return port for server */
        public int getPort()
        {
          return port;
        }
        /**
         * Convert request into a string for easy re-sending.
         */
        public String toString()
        {
          String req="";
          req= method + " " + URI + " " + version + CRLF;
          req + = headers;
          /* This proxy does not support persistent connections */
          req + = "Connection: close" + CRLF;
          req + = CRLF;
          return req;
        }
    }
```

2. HttpResponse.java 文件

```java
/**
     * HttpResponse-Handle HTTP replies
 */
import java.io.*;
import java.net.*;
import java.util.*;
public class HttpResponse
{
    final static String CRLF= "\r\n";
    /** How big is the buffer used for reading the object */
    //final static int BUF_SIZE= 8192;
```

```java
final static int BUF_SIZE= 8192;//
/** Maximum size of objects that this proxy can handle. For the
 * moment set to 100 KB. You can adjust this as needed. */
final static int MAX_OBJECT_SIZE= 100000;
/** Reply status and headers */
String version;
int status;
String statusLine= "";
String headers= "";
/* Body of reply */
byte[] body= new byte[MAX_OBJECT_SIZE];
/** Read response from server.*/
public HttpResponse(DataInputStream/* BufferedReader */fromServer)
{
    /* Length of the object */
    int length= -1;
    boolean gotStatusLine= false;
    /* First read status line and response headers */
try
{
    String line= fromServer.readLine();
    while (line.length()!= 0)
    {
        if (! gotStatusLine)
        {
            statusLine= line;
            gotStatusLine= true;/* 状态行 */
        }
        else
        {
            headers += line + CRLF;/* 首部行 */
        }
        /*Get length of content as indicated by
        *Content-Length header. Unfortunately this is not
        *present in every response. Some servers return the
        *header "Content-Length", others return
        *"Content-length". You need to check for both
        *here. */
        if (line.startsWith("Content-Length")||
```

```java
                line.startsWith("Content-length"))
            {
                String[] tmp= line.split(" ");
                length= Integer.parseInt(tmp[1]);
            }
            line= fromServer.readLine();
        }
        System.out.println(headers);
    }
    catch (IOException e) /**/
    {
        System.out.println("Error reading headers from server: " + e);
        return;
    }
    try
        {
        int bytesRead= 0;
        byte buf[]= new byte[BUF_SIZE];
        boolean loop= false;
        /* If we didn't get Content-Length header, just loop until
         * the connection is closed.*/
        if (length == -1)
        {
            loop= true;
        }
        /* Read the body in chunks of BUF_SIZE and copy the chunk
         * into body. Usually replies come back in smaller chunks
         * than BUF_SIZE. The while-loop ends when either we have
         * read Content-Length bytes or when the connection is
         * closed (when there is no Connection-Length in the
         * response.*/
        while (bytesRead <  length || loop)
        {
            /* Read it in as binary data */
            int res= fromServer.read(buf,bytesRead,buf.length);
            if (res == -1)
            {
                break;
            }
```

```java
        /* Copy the bytes into body. Make sure we don't exceed
         * the maximum object size.*/
            for( int i=0; i< res && (i+ bytesRead)< MAX_OBJECT_SIZE; i++ )
            {
                body[i+ bytesRead]= buf[i];
                }
            bytesRead += res;
            }
        }
        catch (Exception e)
        {
            System.out.println("Error reading response body: " + e);
            return;
        }
    }
    /**
     * Convert response into a string for easy re-sending.
     *
     *
     */
    public String toString()
    {
        String bo_dy= "";
        String res= "";
        res= statusLine + CRLF;
        res += headers;
        res += CRLF;
        try
        {
            bo_dy= new String(body);
        }
        catch (Exception e)
        {
            System.out.println("无法转换格式");
        }
        res+= bo_dy;
        return res;
    }
}
```

3. ProxyCache.java 文件

```java
/**
 * ProxyCache.java-Simple caching proxy
 */
import java.net.*;
import java.io.*;
import java.util.*;
public class ProxyCache
{
    /** Port for the proxy */
    private static int port;
    /** Socket for client connections */
    private static ServerSocket socket;
    /** Create the ProxyCache object and the socket */
    public static void init(int p)
    {
        port= p;
        try
        {
            socket= new ServerSocket(p);
        }
        catch (IOException e)
        {
            System.out.println("Error creating socket: " + e);
            System.exit(-1);
        }
    }
    public static void handle(Socket client)
    {
        Socket server= null;
        HttpRequest request= null;
        HttpResponse response= null;
        /* Process request.  If there are any exceptions, then simply
         * return and end this request.  This unfortunately means the
         * client will hang for a while, until it timeouts.*/
        /* Read request */
        try
```

```
            {
BufferedReader fromClient= new BufferedReader(new InputStreamReader(client.getInputStream
()));
        request= new HttpRequest(fromClient);
        System.out.println("从浏览器接受请求,并构造 REQUEST!");
    }
    catch (IOException e)
    {
        System.out.println("Error reading request from client: " + e);
        return;
    }
    /* Send request to server */
    try
        {
        /* Open socket and write request to socket */
        server= new Socket(request.getHost(),request.getPort());
        DataOutputStream toServer= new DataOutputStream(server.getOutputStream());
        System.out.println("构造 RESPONSE 给 WEB 服务器!");
        System.out.println(request.toString());
        toServer.writeBytes(request.toString());
    }
    catch (UnknownHostException e)
    {
        System.out.println("Unknown host: " + request.getHost());
        System.out.println(e);
        return;
    }
    catch (IOException e)
    {
        System.out.println("Error writing request to server: " + e);
        return;
    }
    /* Read response and forward it to client */
    try
    {
        System.out.println("准备接受 WEB 服务器返回的对象");
DataInputStream fromServer =    new DataInputStream(server.getInputStream());
        response= new HttpResponse(fromServer);
        System.out.println("接收 WEB 服务器返回的对象,并构造成返回给浏览器的
```

```java
RESPONSE");
        DataOutputStream toClient= new DataOutputStream(client.getOutputStream);
            toClient.writeBytes(response.toString().trim());
            FileOutputStream fs= new FileOutputStream("uu.txt");
            fs.write(response.toString().trim().getBytes());
            System.out.println("将构造好的 RESPONSE 对象返回给浏览器!");
            /* Write response to client. First headers, then body */
            client.close();
            server.close();
            System.out.println("返回给浏览器对象结束,并关闭连接!");
            /* Insert object into the cache */
            /* Fill in (optional exercise only) */
        }
        catch (IOException e)
        {
            System.out.println("Error writing response to client: " + e);
            return;
        }
    }
    /** Read command line arguments and start proxy */
    public static void main(String args[])
    {
        int myPort= 0;

        try
        {
            myPort= Integer.parseInt(args[0]);
        }
        catch (ArrayIndexOutOfBoundsException e)
        {
            System.out.println("Need port number as argument");
            System.exit(-1);
        }
        catch (NumberFormatException e)
        {
            System.out.println("Please give port number as integer.");
            System.exit(-1);
        }
        init(myPort);
```

```
/** Main loop. Listen for incoming connections and spawn a new
 * thread for handling them */
Socket client= null;
while (true)
{try
    {client= socket.accept();
        handle(client);
    }
    catch (IOException e)
    {System.out.println("Error reading request from client: " + e);
        /* Definitely cannot continue processing this request,
         * so skip to next iteration of while loop. */
        continue;
    }
  }
 }
}
```

第 12 章　实现一个可靠传输协议

12.1　实 验 目 标

实现一个简单的单向可靠数据传输协议。这个系统有两种版本，分别是比特交替协议版本和 GBN 版本，本章选择的是比特交替协议版本。

12.2　系统设计与组成

本系统包括发送方 A 和接收方 B。

本系统依据网络分层结构在一台 PC 上模拟对数据的发送、传输以及接收并检测。程序主要分三层实现，上层（第五层）负责向下层发送数据；第四层负责可靠数据传输协议部分；下层（第三层）为不可靠信道，负责传输分组，可能会出现丢包、延迟或比特差错。本程序的关键在于第三层的传输媒体的实现方式：双向链表以及三层之间的联系。系统框架如图 12-1 所示。

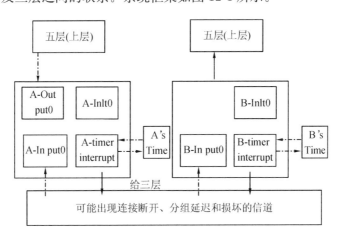

图 12-1　可靠数据传输协议系统框架

实现步骤如下。

第一步，对环境进行初始化，并对发送方 A 和接收方 B 初始化。

第二步，单向数据传输过程实现。

发送方 A 的 FSM 如图 12-2 所示。

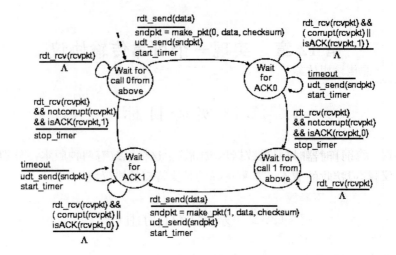

图 12-2　发送方 A 的 FSM 描述

接收方 B 的 FSM 如图 12-3 所示。

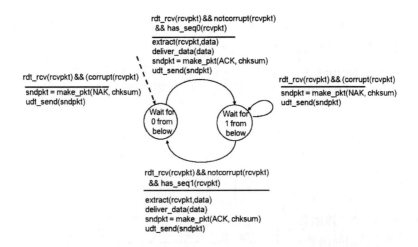

图 12-3　接收方 B 的 FSM 描述

其中一次数据传输过程如图 12-4 所示。

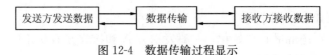

图 12-4　数据传输过程显示

12.3 重要方法

第三层的传输媒体靠一个双向链表来实现，每一个节点均为一个结构体。该链表中记载每次传输分组时按照传输分组的各个事件的时间排序。发送方 A 有一个计时器，以及开启和关闭计时器的两个函数 starttimer（）、stoptimer（）。在链表中，每一次发生事件均创建一个节点（包括开启计时器），开启计时器时创建节点，关闭计时器时删除该节点，所以，最终链表中只包括每一次对数据进行发送、传输或接收的节点。

第四层主要有以下函数（设定 A 为发送方，B 为接收方）：A_output（message），该函数用于接收第 5 层发送来的消息，并把消息信息、设定序号、校验和封装成数据包，通过调用 to_layer3（）函数，将数据包发送到下一层，即第 3 层，并保证数据有序，正确地发送到接收方 B。最后将该事件加入第三层链表中。

A_input（packet），该函数用于接收来自接收方 B 发送的确认信息。即从下层接收到 B 发送回来的确认消息，查看校验和，看是否正确，查看是 ACK 或 NAK；并查看序号是否是该数据包的确认信息。最后将该事件加入第三层链表中。

A_timerinterrupt（），该函数将调用 A 发送方的 starttimer（）和 stoptimer（）函数，用于实现超时事件发生时的中断。最后将该事件加入第三层链表中。

A_init（），该函数用在其他函数之前，用于初始化 A 方的数据。

B_input（packet），该函数用于接收来自发送方 A 发送的数据包。即从下层接收到 B 发送回来的确认消息，查看校验和，看是否正确；并查看序号是否是想要的数据包的序号。最后将该事件加入第三层链表中。

B_init（），该函数用在其他函数之前，用于初始化 B 方的数据。

其中用于调用的函数有四个。

starttimer（），用于开启发送方 A 的计时器。但其实该程序的计时器并非真正的计时，只是设定它的超时时间，以确定是否超时。该函数完成后创建一个节点加入链表中。

stoptimer（），用于停止计时器，在 A 的超时中断中使用。该函数完成后取消在链表中 stattimer（）加入的节点。

to_layer3（），用于将数据包传输到第 3 层，并创建节点，加入链表中。

to_layer5（），用于将消息传送到第 5 层，并创建节点，加入链表中。

要注意的部分如下。

(1) 随机数的生成：依赖函数 jimsrand（ ）生成。
(2) tracing：用于调试方便。
(3) 第 5 层发送数据的速率：尽量大点，推荐用 1000。
(4) 该程序中的事件采用的均不是真正的时间，而是用一个整数值来模拟。

12.4 开发环境

此系统使用 C 语言开发，运行在 Windows 平台下，使用前应安装并配置好相关开发环境。

12.5 运行结果

以下分别为不同条件下的实验结果，如图 12-5～图 12-8 所示。

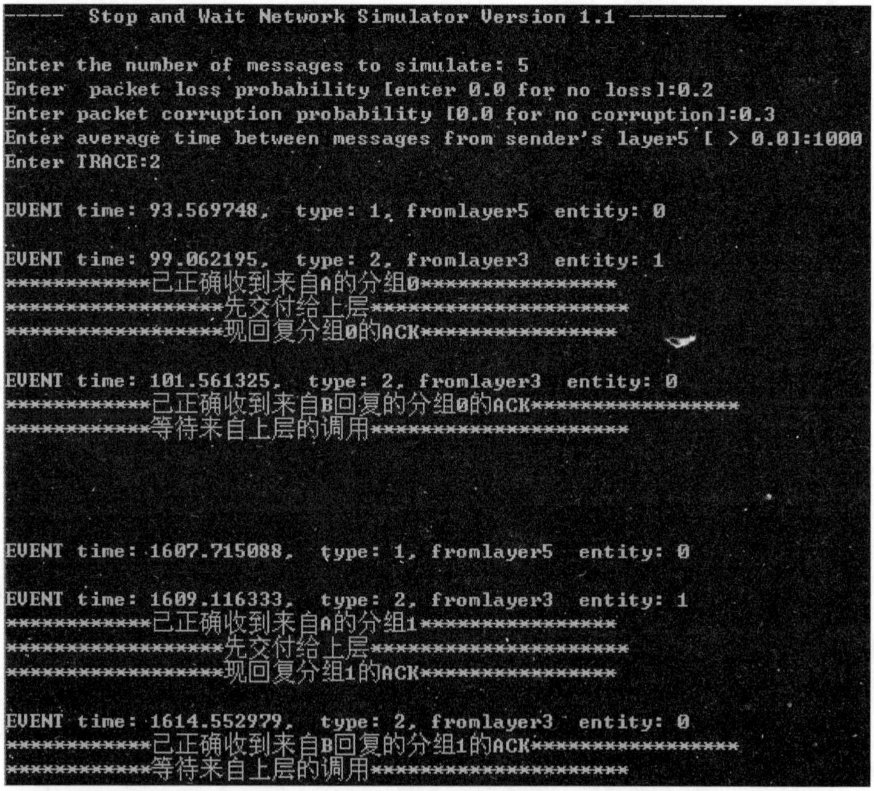

```
------ Stop and Wait Network Simulator Version 1.1 ------
Enter the number of messages to simulate: 5
Enter  packet loss probability [enter 0.0 for no loss]:0.2
Enter packet corruption probability [0.0 for no corruption]:0.3
Enter average time between messages from sender's layer5 [ > 0.0]:1000
Enter TRACE:2

EVENT time: 93.569748, type: 1, fromlayer5  entity: 0

EVENT time: 99.062195, type: 2, fromlayer3  entity: 1
**************已正确收到来自A的分组0*******************
*****************先交付给上层***********************
***************现回复分组0的ACK*********************

EVENT time: 101.561325, type: 2, fromlayer3  entity: 0
**************已正确收到来自B回复的分组0的ACK***************
*****************等待来自上层的调用*********************




EVENT time: 1607.715088, type: 1, fromlayer5  entity: 0

EVENT time: 1609.116333, type: 2, fromlayer3  entity: 1
**************已正确收到来自A的分组1*******************
*****************先交付给上层***********************
***************现回复分组1的ACK*********************

EVENT time: 1614.552979, type: 2, fromlayer3  entity: 0
**************已正确收到来自B回复的分组1的ACK***************
*****************等待来自上层的调用*********************
```

图 12-5　窗口命令显示数据传输①

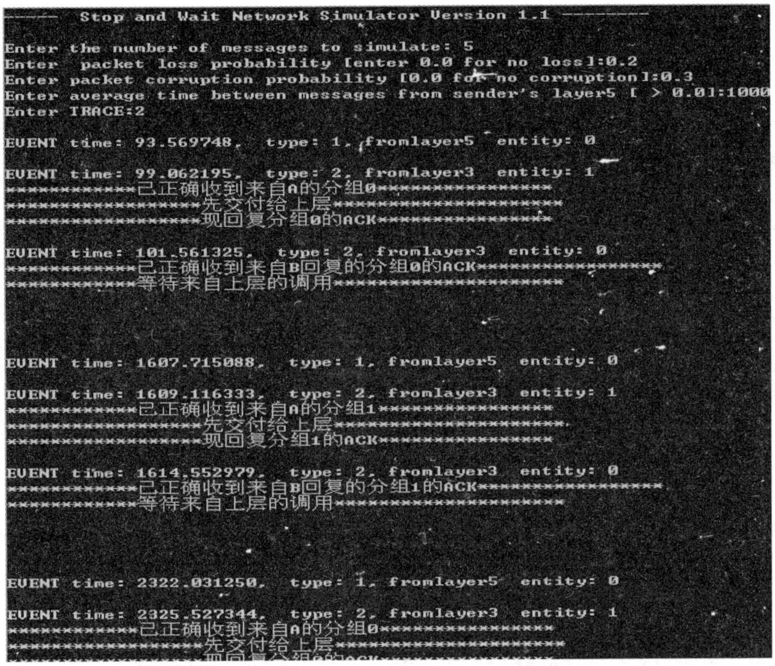

第12章 实现一个可靠传输协议

```
------- Stop and Wait Network Simulator Version 1.1 -------
Enter the number of messages to simulate: 5
Enter  packet loss probability [enter 0.0 for no loss]:0.2
Enter  packet corruption probability [0.0 for no corruption]:0.3
Enter  average time between messages from sender's layer5 [ > 0.0]:1000
Enter TRACE:2
EVENT time: 93.569748,  type: 1, fromlayer5  entity: 0

EVENT time: 99.062195,  type: 2, fromlayer3  entity: 1
***************已正确收到来自A的分组0***************
********************先交付给上层********************
*******************现回复分组0的ACK*****************

EVENT time: 101.561325,  type: 2, fromlayer3  entity: 0
************已正确收到来自B回复的分组0的ACK**********
*******************等待来自上层的调用***************



EVENT time: 1607.715088,  type: 1, fromlayer5  entity: 0
EVENT time: 1609.116333,  type: 2, fromlayer3  entity: 1
***************已正确收到来自A的分组1***************
********************先交付给上层********************
*******************现回复分组1的ACK*****************

EVENT time: 1614.552979,  type: 2, fromlayer3  entity: 0
************已正确收到来自B回复的分组1的ACK**********
*******************等待来自上层的调用***************



EVENT time: 2322.031250,  type: 1, fromlayer5  entity: 0
EVENT time: 2325.527344,  type: 2, fromlayer3  entity: 1
***************已正确收到来自A的分组0***************
********************先交付给上层********************
```

图 12-6　窗口命令显示数据传输②

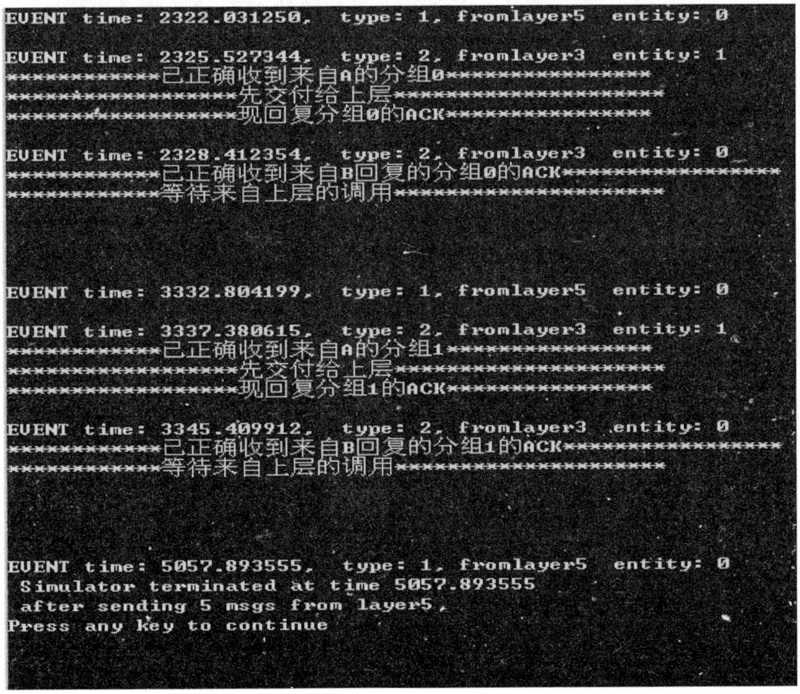

```
EVENT time: 2322.031250,  type: 1, fromlayer5  entity: 0
EVENT time: 2325.527344,  type: 2, fromlayer3  entity: 1
***************已正确收到来自A的分组0***************
********************先交付给上层********************
*******************现回复分组0的ACK*****************

EVENT time: 2328.412354,  type: 2, fromlayer3  entity: 0
************已正确收到来自B回复的分组0的ACK**********
*******************等待来自上层的调用***************



EVENT time: 3332.804199,  type: 1, fromlayer5  entity: 0
EVENT time: 3337.380615,  type: 2, fromlayer3  entity: 1
***************已正确收到来自A的分组1***************
********************先交付给上层********************
*******************现回复分组1的ACK*****************

EVENT time: 3345.409912,  type: 2, fromlayer3  entity: 0
************已正确收到来自B回复的分组1的ACK**********
*******************等待来自上层的调用***************



EVENT time: 5057.893555,  type: 1, fromlayer5  entity: 0
Simulator terminated at time 5057.893555
 after sending 5 msgs from layer5.
Press any key to continue
```

图 12-7　窗口命令显示数据传输③

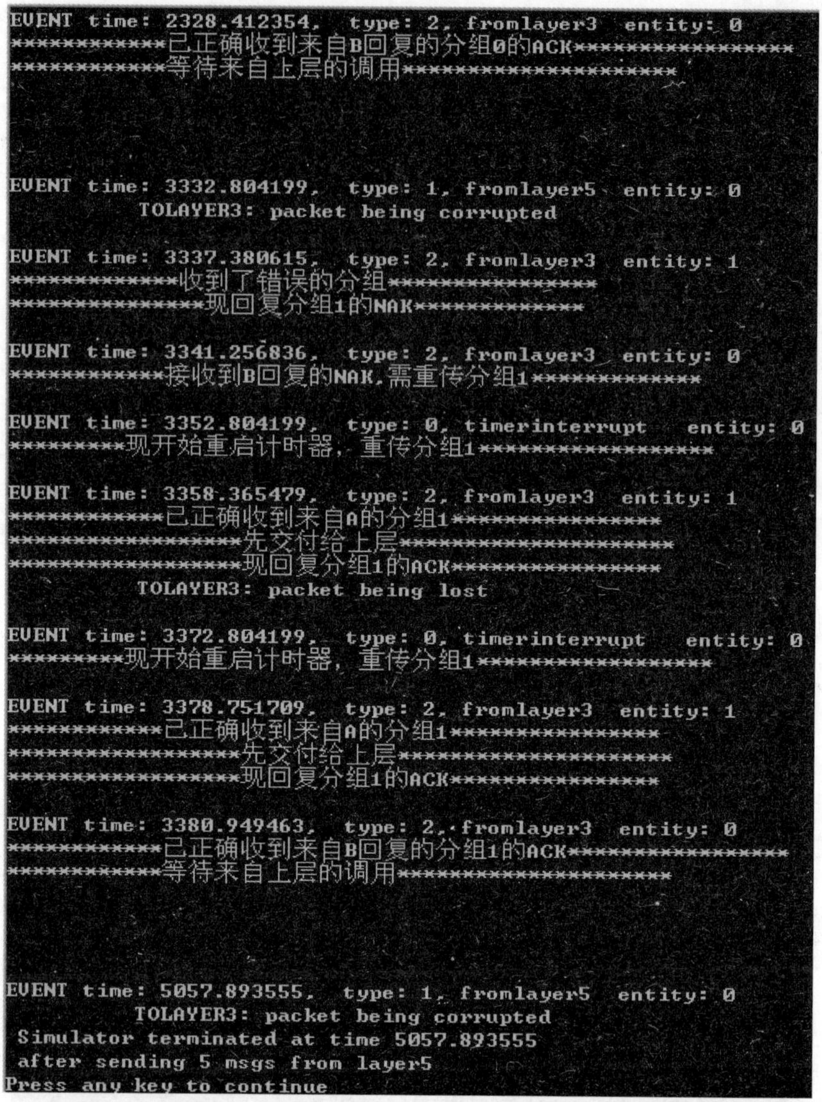

```
EVENT time: 2328.412354,  type: 2, fromlayer3  entity: 0
***************已正确收到来自B回复的分组0的ACK****************
***************等待来自上层的调用****************

EVENT time: 3332.804199,  type: 1, fromlayer5  entity: 0
         TOLAYER3: packet being corrupted

EVENT time: 3337.380615,  type: 2, fromlayer3  entity: 1
***************收到了错误的分组****************
***************现回复分组1的NAK****************

EVENT time: 3341.256836,  type: 2, fromlayer3  entity: 0
***************接收到B回复的NAK,需重传分组1****************

EVENT time: 3352.804199,  type: 0, timerinterrupt  entity: 0
***********现开始重启计时器,重传分组1****************

EVENT time: 3358.365479,  type: 2, fromlayer3  entity: 1
***************已正确收到来自A的分组1****************
***************先交付给上层****************
***************现回复分组1的ACK****************
         TOLAYER3: packet being lost

EVENT time: 3372.804199,  type: 0, timerinterrupt  entity: 0
***********现开始重启计时器,重传分组1****************

EVENT time: 3378.751709,  type: 2, fromlayer3  entity: 1
***************已正确收到来自A的分组1****************
***************先交付给上层****************
***************现回复分组1的ACK****************

EVENT time: 3380.949463,  type: 2, fromlayer3  entity: 0
***************已正确收到来自B回复的分组1的ACK****************
***************等待来自上层的调用****************

EVENT time: 5057.893555,  type: 1, fromlayer5  entity: 0
         TOLAYER3: packet being corrupted
Simulator terminated at time 5057.893555
after sending 5 msgs from layer5
Press any key to continue
```

图 12-8　窗口命令显示数据传输④

12.6　源　代　码

Rdt.c 文件源代码如下。

```
#include <stdio.h>
/***************************************************************
```

ALTERNATING BIT AND GO- BACK- N NETWORK EMULATOR: VERSION 1.1   J. F. Kurose
This code should be used for PA2, unidirectional or bidirectional data transfer protocols
(from A to B. Bidirectional transfer of data is for extra credit and is not required).
Network properties:
-one way network delay averages five time units (longer if there are other messages in the channel for GBN), but can be larger
-packets can be corrupted (either the header or the data portion) or lost, according to user-defined probabilities
-packets will be delivered in the order in which they were sent (although some can be lost).
**/
#define BIDIRECTIONAL 0   /* change to 1 if you're doing extra credit */
                              /* and write a routine called B_output */
#define   TIMER_INTERRUPT 0
#define   FROM_LAYER5   1
#define   FROM_LAYER3   2
#define   OFF   0
#define   ON   1
#define   A   0
#define   B   1
int A_againsum;   //重传一个分组的总次数
/* a "msg" is the data unit passed from layer 5 (teachers code) to layer   */
/* 4 (students' code). It contains the data (characters) to be delivered */
/* to layer 5 via the students transport level protocol entities.   */
struct msg {
   char data[20];
   };
/* a packet is the data unit passed from layer 4 (students code) to layer */
/* 3 (teachers code). Note the pre-defined packet structure, which all   */
/* students must follow.*/
struct pkt {
   int seqnum;
   int acknum;
   int checksum;
   char payload[20];
   };
struct   pkt   A_pkt;//内存中的旧分组
struct   pkt   ackpkt;//B回复的ACK分组或NAK分组
/******** STUDENTS WRITE THE NEXT SEVEN ROUTINES ********/
/* checksum 判断分组是否正确 */

```c
int check(packet)
struct pkt packet;
{
    int i,newchecksum;
    newchecksum= packet. seqnum+ packet. acknum;
    for(i= 0;i< 20;i++ )
        newchecksum + = packet. payload[i];
    if(newchecksum= = packet. checksum)
        return 1;
    else
        return 0;
}
/* called from layer 5, passed the data to be sent to other side */
void A_output(message)
struct msg message;
{
    int i;
    A_pkt. acknum= 0;
    A_pkt. checksum= A_pkt. seqnum+ A_pkt. acknum;
    for(i= 0;i< 20;i+ + )
    {
        A_pkt. payload[i]= message. data[i];
        A_pkt. checksum+ = A_pkt. payload[i];
    }
    starttimer(A,20. 0);
    tolayer3(A,A_pkt);
}
void B_output(message)/* need be completed only for extra credit */
struct msg message;
{
}
/* called from layer 3, when a packet arrives for layer 4 */
void A_input(packet)
struct pkt packet;
{
    if(check(packet))
    {
        if(packet. acknum= = 1&&A_pkt. seqnum= = packet. seqnum)
        {
```

```
                printf("*********** 已正确收到来自B回复的分组%d的ACK ********
                ********\n",packet.seqnum);
                stoptimer(A);
                A_pkt.seqnum= 1-A_pkt.seqnum;
            printf("*********** 等待来自上层的调用 ******************** \n\n\n\n");
            }
            else if(packet.acknum= = 0&&A_pkt.seqnum= = packet.seqnum)
                    printf("*********** 接收到B回复的NAK,需重传分组%d ******
                    ****** \n",packet.seqnum);
                else
                printf("*********** 接收到失序的分组,丢弃 ******************* \n");
}
else
    printf("*********** 接收到损坏的分组,丢弃 ************** \n");
}
/* called when A's timer goes off */
void A_timerinterrupt()
{
    if(A_againsum< 3)
    {
        printf("********* 现开始重启计时器,重传分组%d ***************** \n",A_
        pkt.seqnum);
        A_againsum+ + ;
        starttimer(A,20.0);
        tolayer3(A,A_pkt);
    }
    else
    {
        printf("********* 重传的次数过多,丢弃分组,等待上层的调用 ***********
        *** \n");
        A_againsum= 0;
        A_pkt.seqnum= 1-A_pkt.seqnum;
    }
}
/* the following routine will be called once (only) before any other */
/* entity A routines are called. You can use it to do any initialization */
void A_init()
{
    A_againsum= 0;
```

```
   A_pkt.seqnum= 0;
}
/* Note that with simplex transfer from a-to-B, there is no B_output() */


/* called from layer 3, when a packet arrives for layer 4 at B */
void B_input(packet)
struct pkt packet;
{
   int i;
   struct msg message;
   if(check(packet))
   {
   for(i= 0;i< 20;i+ + )
             message.data[i]= packet.payload[i];
   printf("*********** 已正确收到来自 A 的分组%d ************\n",
   packet.seqnum);
   printf("***************** 先交付给上层 ******************\n");
   tolayer5(B,message.data);
   printf("**************** 现回复分组%d 的 ACK ***************\n",
   packet.seqnum);
   ackpkt.acknum= 1;
   }
   else
   {
      printf("*********** 收到了错误的分组 ***************\n");
      printf("************** 现回复分组%d 的 NAK ************\n",packet.seqnum);
      ackpkt.acknum= 0;
   }
   ackpkt.seqnum= packet.seqnum;
   ackpkt.checksum= ackpkt.seqnum+ ackpkt.acknum;
   tolayer3(B,ackpkt);
}
/* called when B's timer goes off */
void B_timerinterrupt()
{
   printf("B_timerinterrupt!\n");
}
/* the following rouytine will be called once (only) before any other */
/* entity B routines are called. You can use it to do any initialization */
```

```
void B_init()
{
    int i;
    for(i= 0;i< 20;i+ + )
    {
        ackpkt.payload[i]= 0;
    }
}
/***************************************************************
*************** NETWORK EMULATION CODE STARTS BELOW **********
The code below emulates the layer 3 and below network environment:
-emulates the tranmission and delivery (possibly with bit- level corruption and packet
loss) of packets across the layer 3/4 interface
-handles the starting/stopping of a timer, and generates timer interrupts (resulting in
calling students timer handler).
-generates message to be sent (passed from later 5 to 4)

THERE IS NOT REASON THAT ANY STUDENT SHOULD HAVE TO READ OR UNDERSTAND
THE CODE BELOW. YOU SHOULD NOT TOUCH, OR REFERENCE (in your code) ANY OF THE DATA
STRUCTURES BELOW. If you're interested in how I designed the emulator, you're welcome to
look at the code-but again, you should have to, and you defeinitely should not have
to modify
***************************************************************/
struct event {
    float evtime;           /* event time */
    int evtype;             /* event type code(3,5 或中断) */
    int eventity;           /* entity where event occurs(A 或 B) */
    struct pkt * pktptr;    /* ptr to packet (if any) assoc w/ this event */
    struct event * prev;
    struct event * next;
};
struct event * evlist= NULL;   /* the event list */
/* possible events: */
int TRACE= 1;              /* for my debugging */
int nsim= 0;               /* number of messages from 5 to 4 so far */
int nsimmax= 0;            /* number of msgs to generate, then stop */
float time= 0.000;
float lossprob;            /* probability that a packet is dropped */
float corruptprob;         /* probability that one bit is packet is flipped */
```

```c
        float lambda;                   /* arrival rate of messages from layer 5 */
        int   ntolayer3;                /* number sent into layer 3 */
        int   nlost;                    /* number lost in media */
        int ncorrupt;                   /* number corrupted by media */
        main()
        {
          struct event *eventptr;
          struct msg   msg2give;
          struct pkt   pkt2give;
          int i,j;
          char c;
          init();
          A_init();
          B_init();
          while (nsim< nsimmax) {
               eventptr= evlist;             /* get next event to simulate */
               if (eventptr= = NULL)
                    goto terminate;
               evlist= evlist-> next;        /* remove this event from event list */
               if (evlist! = NULL)
                    evlist-> prev= NULL;
               if (TRACE> = 2) {
                    printf("\nEVENT time: % f",eventptr-> evtime);
    printf(" type: % d",eventptr-> evtype);
    if (eventptr-> evtype= = 0)
         printf(", timerinterrupt ");
              else if (eventptr-> evtype= = 1)
                   printf(", fromlayer5 ");
              else
    printf(", fromlayer3 ");
              printf(" entity: % d\n",eventptr-> eventity);
              }
          time= eventptr-> evtime;            /* update time to next event time */
//        if(nsim< nsimmax) break;
          if (eventptr-> evtype = = FROM_LAYER5 ) {
              generate_next_arrival();  /* set up future arrival */
              /* fill in msg to give with string of same letter */
  j= nsim % 26;
              for (i= 0; i< 20; i+ + )
```

```
              msg2give.data[i]= 97 + j;
                 if (TRACE> 2) {
                    printf("          MAINLOOP: data given to student: ");
for (i= 0; i< 20; i+ + )
                       printf("% c", msg2give.data[i]);
printf("\n");
                 }
                 nsim+ + ;
                 if (eventptr- > eventity = = A)
                    A_output(msg2give);
                  else
                    B_output(msg2give);
              }
           else if (eventptr- > evtype = =    FROM_LAYER3) {
              pkt2give.seqnum= eventptr- > pktptr- > seqnum;
              pkt2give.acknum= eventptr- > pktptr- > acknum;
              pkt2give.checksum= eventptr- > pktptr- > checksum;
              for (i= 0; i< 20; i+ + )
                 pkt2give.payload[i]= eventptr- > pktptr- > payload[i];
    if (eventptr- > eventity = = A)       /* deliver packet by calling */
       A_input(pkt2give);            /* appropriate entity */
              else
       B_input(pkt2give);
    free(eventptr- > pktptr);            /* free the memory for packet */
                 }
              else if (eventptr- > evtype = =    TIMER_INTERRUPT) {
                 if (eventptr- > eventity = = A)
    A_timerinterrupt();
                  else
    B_timerinterrupt();
                 }
                 else {
           printf("INTERNAL PANIC: unknown event type \n");
                 }
        free(eventptr);
              }
terminate:
    printf(" Simulator terminated at time % f\n after sending % d msgs from layer5\n",time,
    nsimmax);
```

```c
}
init()                              /* initialize the simulator */
{
    int i;
    float sum, avg;
    float jimsrand();
    printf("-----Stop and Wait Network Simulator Version 1.1 --------\n\n");
    printf("Enter the number of messages to simulate: ");
    scanf("%d",&nsimmax);
    printf("Enter packet loss probability [enter 0.0 for no loss]:");
    scanf("%f",&lossprob);
    printf("Enter packet corruption probability [0.0 for no corruption]:");
    scanf("%f",&corruptprob);
    printf("Enter average time between messages from sender's layer5 [ > 0.0]:");
    scanf("%f",&lambda);
    printf("Enter TRACE:");
    scanf("%d",&TRACE);
    srand(9999);                    /* init random number generator */
    sum= 0.0;                       /* test random number generator for students */
    for (i= 0; i< 1000; i++)
        sum= sum+ jimsrand();       /* jimsrand() should be uniform in [0,1] */
    avg= sum/1000.0;
    if (avg < 0.25 || avg > 0.75) {
        printf("It is likely that random number generation on your machine\n" );
        printf("is different from what this emulator expects.  Please take\n");
        printf("a look at the routine jimsrand() in the emulator code. Sorry. \n");
        exit();
    }
    ntolayer3= 0;
    nlost= 0;
    ncorrupt= 0;
    time= 0.0;                      /* initialize time to 0.0 */
    generate_next_arrival();        /* initialize event list */
}
/********************************************************************/
/* jimsrand(): return a float in range [0,1].  The routine below is used to */
/* isolate all random number generation in one location.  We assume that the */
/* system-supplied rand() function return an int in therange [0,mmm]       */
/********************************************************************/
```

```
float jimsrand()
{
    double mmm= 32767;   /* 2147483647;  /* largest int  -  MACHINE DEPENDENT!!!!!!!!  */
    float x;                             /* individual students may need to change mmm */
    x= rand()/mmm;                       /* x should be uniform in [0,1] */
    return(x);
}
/******************EVENT HANDLINE ROUTINES ******/
/*    The next set of routines handle the event list   */
/**********************************************/
generate_next_arrival()
{
    double x,log(),ceil();
    struct event *evptr;
      char *malloc();
    float ttime;
    int tempint;
    if (TRACE> 2)
        printf("          GENERATE NEXT ARRIVAL: creating new arrival\n");
    x= lambda * jimsrand() * 2;  /* x is uniform on [0,2 * lambda] */
                                 /* having mean of lambda         */
    evptr= (struct event *)malloc(sizeof(struct event));
    evptr-> evtime= time + x;
    evptr-> evtype =  FROM_LAYER5;
    if (BIDIRECTIONAL && (jimsrand()> 0.5))
        evptr-> eventity= B;
      else
        evptr-> eventity= A;
    insertevent(evptr);
}
insertevent(p)
    struct event *p;
{
    struct event *q,*qold;
    if (TRACE> 2) {
        printf("          本事件时间: time is % lf\n",time);
        printf("          INSERTEVENT: future time will be % lf\n",p-> evtime);
        }
    q= evlist;      /* q points to header of list in which p struct inserted */
```

```
        if (q== NULL) {    /* list is empty */
            evlist= p;
            p-> next= NULL;
            p-> prev= NULL;
            }
        else {
            for (qold= q; q ! = NULL && p-> evtime >  q-> evtime; q= q-> next)
                qold= q;
            if (q== NULL) {   /* end of list */
                qold-> next= p;
                p-> prev= qold;
                p-> next= NULL;
                }
              else if (q== evlist) { /* front of list */
                p-> next= evlist;
                p-> prev= NULL;
                p-> next-> prev= p;
                evlist= p;
                }
              else {      /* middle of list */
                p-> next= q;
                p-> prev= q-> prev;
                q-> prev-> next= p;
                q-> prev= p;
                }
            }
}
printevlist()
{
   struct event * q;
   int i;
   printf("- - - - - - - - - - - - \nEvent List Follows:\n");
   for(q= evlist; q! = NULL; q= q-> next) {
      printf("Event time: % f, type: % d entity: % d\n",q-> evtime,q-> evtype,q-> even-
      tity);
      }
   printf("- - - - - - - - - - - - \n");
}
/****************** Student- callable ROUTINES ********************/
```

```
/* called by students routine to cancel a previously- started timer */
stoptimer(AorB)
int AorB;   /* A or B is trying to stop timer */
{
struct event *q, *qold;
if (TRACE> 2)
    printf("          STOP TIMER: stopping timer at % f\n",time);
/* for (q= evlist; q! = NULL && q- > next! = NULL; q= q- > next)   */
    for (q= evlist; q! = NULL ; q= q- > next)
        if ( (q- > evtype= = TIMER_INTERRUPT  && q- > eventity= = AorB) ) {
            /* remove this event */
            if (q- > next= = NULL && q- > prev= = NULL)
                evlist= NULL;           /* remove first and only event on list */
            else if (q- > next= = NULL) /* end of list - there is one in front */
                q- > prev- > next= NULL;
            else if (q= = evlist) { /* front of list - there must be event after */
                q- > next- > prev= NULL;
                evlist= q- > next;
            }
            else {    /* middle of list */
                q- > next- > prev= q- > prev;
                q- > prev- > next =  q- > next;
            }
        free(q);
        return;
        }
    printf("Warning: unable to cancel your timer. It wasn't running. \n");
}
starttimer(AorB,increment)
int AorB;   /* A or B is trying to start timer */
float increment;
{
struct event *q;
struct event *evptr;
char * malloc();
if (TRACE> 2)
    printf("          START TIMER: starting timer at % f\n",time);
/* be nice: check to see if timer is already started, if so, then  warn */
```

```c
    /* for (q= evlist; q!=NULL && q->next!=NULL; q= q->next) */
        for (q= evlist; q!=NULL ; q= q->next)
            if ((q->evtype== TIMER_INTERRUPT  && q->eventity== AorB)) {
                printf("Warning: attempt to start a timer that is already started\n");
                return;
            }
    /* create future event for when timer goes off */
        evptr= (struct event *)malloc(sizeof(struct event));
        evptr->evtime =   time + increment;
        evptr->evtype =   TIMER_INTERRUPT;
        evptr->eventity= AorB;
        insertevent(evptr);
}
/*********************** TOLAYER3 ************* */
tolayer3(AorB,packet)
int AorB;    /* A or B is trying to stop timer */
struct pkt packet;
{
struct pkt * mypktptr;
struct event * evptr, * q;
char * malloc();
float lastime, x, jimsrand();
int i;
ntolayer3++;
/* simulate losses: */
if (jimsrand() < lossprob) {
        nlost++;
        if (TRACE> 0)
printf("          TOLAYER3: packet being lost\n");
        return;
    }
/* make a copy of the packet student just gave me since he/she may decide */
/* to do something with the packet after we return back to him/her */
mypktptr= (struct pkt *)malloc(sizeof(struct pkt));
mypktptr->seqnum= packet.seqnum;
mypktptr->acknum= packet.acknum;
mypktptr->checksum= packet.checksum;
for (i= 0; i< 20; i++)
     mypktptr->payload[i]= packet.payload[i];
```

```
   if (TRACE> 2)  {
      printf("          TOLAYER3: seq: % d, ack % d, check: % d ", mypktptr- > seqnum,
   mypktptr- > acknum,  mypktptr- > checksum);
         for (i= 0; i< 20; i+ + )
            printf("% c",mypktptr- > payload[i]);
         printf("\n");
      }
/* create future event for arrival of packet at the other side */
   evptr= (struct event * )malloc(sizeof(struct event));
   evptr- > evtype= FROM_LAYER3; /* packet will pop out from layer3 */
   evptr- > eventity= (AorB+ 1)% 2; /* event occurs at other entity */
   evptr- > pktptr= mypktptr;         /* save ptr to my copy of packet */
/* finally, compute the arrival time of packet at the other end.
    medium can not reorder, so make sure packet arrives between 1 and 10
    time units after the latest arrival time of packets
    currently in the medium on their way to the destination */
lastime= time;
/* for (q= evlist; q! = NULL && q- > next! = NULL; q= q- > next) */
for (q= evlist; q! = NULL ; q= q- > next)
     if ( (q- > evtype= = FROM_LAYER3  && q- > eventity= = evptr- > eventity) )
        lastime= q- > evtime;
evptr- > evtime =   lastime + 1 + 9 * jimsrand();
/* simulate corruption: */
if (jimsrand()< corruptprob) {
     ncorrupt+ + ;
     if ( (x= jimsrand()) < .75)
        mypktptr- > payload[0]= 'Z';   /* corrupt payload */
        else if (x < .875)
        mypktptr- > seqnum= 999999;
        else
        mypktptr- > acknum= 999999;
     if (TRACE> 0)
printf("         TOLAYER3: packet being corrupted\n");
     }
  if (TRACE> 2)
     printf("          TOLAYER3: scheduling arrival on other side\n");
insertevent(evptr);
}
tolayer5(AorB,datasent)
```

```
int AorB;
char datasent[20];
{
   int i;
   if (TRACE> 2) {
      printf("          TOLAYER5: data received: ");
      for (i= 0; i< 20; i+ + )
         printf("% c",datasent[i]);
      printf("\n");
   }
}
```

第 13 章 一个分布式异步距离向量算法

13.1 实验目标

设计一段模拟程序,为图 13-1 所示的网络实现一个分布式异步距离向量选路算法。

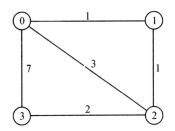

图 13-1 分布式距离算法设计所需网络拓扑

13.2 系统设计与组成

本系统主要针对图 13-1 进行实现,所以包括四个节点,此外,此系统包括三个源文件。

DVsim.h:包含各种定义的头文件。

DVsim.cpp:程序运行的特定虚拟环境,其中有对时间的控制,在某一时间时刻,节点发送距离向量,判断后更新自己的距离表。

Node.cpp:节点根据具体算法对自身的距离表进行更新。

13.3 系统设计

依据距离向量路由算法的基本思想,每个节点不时地向它的每个邻居发送它的距离向量信息,当节点 x 从它的任何一个邻居 v 收到一个新距离向量时,保存 v 的距离向量,然后使用 Bellman-Ford 方程更新它自己的距离向量

$$D_x(y) = \min\{c(x, v) + D_v(y)\}$$

如果节点 x 的距离向量因这个更新步骤而改变,则节点 x 将向它的每个邻居发送它的更新的距离向量。

实现步骤如下：

第一步：定义数组来存放路径长度。

第二步：分别初始化节点 D_0、D_1、D_2、D_3 的距离向量，此时用到以下 4 个子函数，即 void rtinit0（ ）、void rtinit1（ ）、void rtinit2（ ）、void rtinit3（ ）。

第三步：初始化之后，在系统所设置的特定的环境里，分别对各自的距离向量进行更新，此时用到以下 4 个子函数，即 void rtupdate0（struct rtpkt *）、void rtupdate1（struct rtpkt *）、void rtupdate2（struct rtpkt *）、void rtupdate3（struct rtpkt *）。

第四步：用矩阵的形式将结果输出。

13.4 重 要 方 法

rtinit0（ ），模拟一旦开始，该程序就被调用。rtinit（ ）无参数。它应该初始化节点 0 中的距离表，以反映到达节点 1、2、3 的直接费用分别为 1、3、7。在图 13-1 中，所有链路都是双向的，两个方向的费用都相同。在初始化距离表和节点 0 的程序所需的其他数据结构后，它应向其直接连接的邻居（这里为 1、2、3）发送它到其他网络节点的最低费用路径的费用信息。这种最低费用信息通过调用 tolayer2（ ），在一个选路更新分组中发送给相邻节点［其他的 rtinit1（ ）、rtinit2（ ）、rtinit3（ ）与此类似］。

rtudate0（struct rtpkt，*rcvpkt），当节点 0 收到一个由直接相连邻居发给它的选路分组时，调用该程序。参数 *rcvpkt 是一个指向接受分组的指针。rtudate0（ ）是距离向量算法的"核心"，它从其他节点 i 接收的选路更新分组中，包含有节点 i 到所有其他网络节点的当前最短路径费用值。rtudate0（ ）使用这些收到的值来更新其自身的距离表（这是距离向量算法所规定的，即依据 Bellman-Ford 方程）。如果它自己到另外节点的最低费用因此更新而发生改变，则节点 0 通过发送一个选路分组，来通知其直接相连邻居这种最低费用的变化。仅有直接相连的节点才交换选路信息，即节点 1 和 2 相互通信，但节点 1 和 3 将不相互通信［其他的 rtudate1（struct rtpkt，*rcvpkt）、rtudate2（struct rtpkt，*rcvpkt）、rtudate03（struct rtpkt，*rcvpkt）与此类似］。

13.5 开 发 环 境

此系统使用 C 语言开发，运行在 Windows 平台下，使用前应安装并配置好相关开发环境。

13.6 运行结果

初始化节点 D_0、D_1、D_2、D_3 的距离向量后，结果如图 13-2。

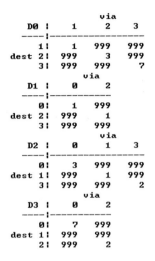

图 13-2 初始化距离向量

矩阵表示 D_i 经过行标（via）到达列标（dest）的路程长度，$i=1, 2, 3, 4$（若此两点之间没有直接到达的路，则表示为 999）。

分别对各自的距离向量进行更新（如节点 1 告诉节点 0 它自己到其他节点的距离向量），正确输出及简单说明如图 13-3 和图 13-4 所示。

> 节点1距离节点0为1，到自己为0，距离节点3为1，距离节点4无

图 13-3 距离向量说明

```
MAIN: rcv event, t=1270.000, at 0 src: 1, dest: 0, contents:  1   0   1 999
                via
     D0 !    1      2     3
     ----:------------------
        1!   1    999   999
     dest 2!   2      3   999
        3! 999    999     7
```

图 13-4 距离向量说明

即节点 0 更新自己到节点 2 的路径长度

$$D_0(2) = \min\{c(0, 1) + D_1(2)\} = 1 + 1 = 2$$

同理，节点 2 在某一时刻向节点 0 发送向量（3, 1, 0, 2），节点 0 根据此向量更新自己的矩阵，如图 13-5 所示。

```
MAIN: rcv event, t=19694.000, at 0 src: 2, dest: 0, contents:    3    1    0    2
                       via
         D0 |    1     2    3
         ---|----------------
           1|    1     4   999
    dest  2|    2     3   999
           3|  999     5     7
```

图 13-5 距离向量更新

$$D_0(1) = \min\{c(0, 2) + D_2(1)\} = 3 + 1 = 4$$

同理

$$D_0(3) = \min\{c(0, 2) + D_2(3)\} = 3 + 2 = 5$$

用矩阵的形式将结果（以 D_0 为例）输出如图 13-6 所示。

```
void printdt0 (struct distance_table *dtptr)
{
  printf("                  via      \n");
  printf("       D0 |    1     2    3 \n");
  printf("       ---|----------------\n");
  printf("         1|  %3d   %3d   %3d\n",dtptr->costs[1][1],dtptr->costs[1][2],dtptr->costs[1][3]);
  printf("  dest  2|  %3d   %3d   %3d\n",dtptr->costs[2][1],dtptr->costs[2][2],dtptr->costs[2][3]);
  printf("         3|  %3d   %3d   %3d\n",dtptr->costs[3][1],dtptr->costs[3][2],dtptr->costs[3][3]);
}
```

图 13-6 距离向量输出

13.7 源 代 码

1. DVsim.h 文件

```
#include    <stdio.h>
#include    <stdlib.h>
#include    <process.h>
#include    <io.h>
#include    <iostream.h>
#define     INF      999
#define     NODES    4
#define     LINKCHANGES   1
/*----------------------------------------------------------
 *    distance_table definition:
 *
 *    costs[i][j]= path cost from this node to node i
 *             when first hop is node j
 *---------------------------------------------------------*/
struct    distance_table
```

```c
    {
        int costs[NODES][NODES];
    };
#ifdef DVSIM_H
/*----------------------------
 *    Trace level
 *--------------------------- */
int    TRACE= 1;
/*------------------------------------
 *    distance tables
 *    minimum path costs
 *    link costs
 *----------------------------------- */
struct    distance_table    dt[NODES];
int    min_cost[NODES][NODES];
int    linkcost[NODES][NODES]= {
        {0,1,3,7},
        {1,0,1,INF},
        {3,1, 0,2},
        {7,INF,2,0},
};
#else
extern int        TRACE;
extern struct     distance_table dt[NODES];
extern int        min_cost[NODES][NODES];
extern int        linkcost[NODES][NODES];
extern int rpc0= 0;
extern int rpc1= 0;
extern int rpc2= 0;
extern int rpc3= 0;
extern int spc0= 0;
extern int spc1= 0;
extern int spc2= 0;
extern int spc3= 0;
#endif
/*----------------------------------------------------------
 *    rtpkt          Routing packet
 *
 *    sourceid       source router
```

```
 *     destid          dest router (must be immediate neighbor)
 *     mincost[]       minimum cost to other nodes
 *-------------------------------------------------------------- */
struct rtpkt {
        int sourceid;
        int destid;
        int mincost[NODES];
};
/*-----------------------------
 *     function prototypes
 *----------------------------- */
void       rtinit0();
void       rtinit1();
void       rtinit2();
void       rtinit3();
void       rtupdate0 (struct rtpkt *);
void       rtupdate1 (struct rtpkt *);
void       rtupdate2 (struct rtpkt *);
void       rtupdate3 (struct rtpkt *);
void       printdt0 (struct distance_table *);
void       printdt1 (struct distance_table *);
void       printdt2 (struct distance_table *);
void       printdt3 (struct distance_table *);
void       linkhandler0 (int, int);
void       linkhandler1 (int, int);
double     get_clock();
void       tolayer2 (struct rtpkt);
/*   END DVsim.h ------------------------------------------- */
```

2. Node.cpp 文件

```
#include     "DVsim.h"
/*-----------------------
 *     rtinit0
 *----------------------- */
void rtinit0()
{
    for(int i=0; i< 4; i++)
        for(int j=0; j< 4; j++)
```

```
            dt[0].costs[i][j]= 999;        //The first step of initializing node0's
                                             costs table.

    for(i= 0; i < 4; i+ + )                //The first step of initializing node0's min_
                                             cost table
        min_cost[0][i]= 999;               //The reason of seperating the initiazing
                                             process is
                                                //just for easy- coding.

    dt[0].costs[0][0]= min_cost[0][0]= 0;//The second step of initializing node0's min_
                                             cost table and
                                           //costs table. Their relationship can be
                                             drawn from the analyzing graph
    dt[0].costs[1][1]= min_cost[0][1]= 1;
    dt[0].costs[2][2]= min_cost[0][2]= 3;
    dt[0].costs[3][3]= min_cost[0][3]= 7;
rtpkt pki;
    pki.sourceid= 0;
    for(i= 0; i < NODES; i+ + )
pki.mincost[i]= dt[0].costs[i][i];//initializing the data segment in the packet.
    pki.destid= 1;
    tolayer2(pki);
    pki.destid= 2;
    tolayer2(pki);
    pki.destid= 3;
    tolayer2(pki);
spc0 + = 3;
}
/*----------------------
 *    rtinit1
 *---------------------- */
void rtinit1()
{
    int i;
for(i= 0; i < NODES; i+ + )
    for(int j= 0; j < NODES; j+ + )
dt[1].costs[i][j]= INF;
for(i= 0; i < NODES; i+ + )
min_cost[1][i]= INF;
```

```
    dt[1].costs[0][0] = min_cost[1][0] = 1;
    dt[1].costs[1][1] = min_cost[1][1] = 0;
    dt[1].costs[2][2] = min_cost[1][2] = 1;
rtpkt pki;
    pki.sourceid = 1;
    for(i = 0; i < NODES; i++)
pki.mincost[i] = dt[1].costs[i][i];
    pki.destid = 0;
    tolayer2(pki);
    pki.destid = 2;
    tolayer2(pki);
    spc1 += 2;
}
/*-----------------------
 *   rtinit2
 *----------------------- */
void rtinit2()
{
    int i;
for(i = 0; i < NODES; i++)
        for(int j = 0; j < NODES; j++)
dt[2].costs[i][j] = INF;
for(i = 0; i < NODES; i++)
min_cost[2][i] = INF;
    dt[2].costs[0][0] = min_cost[2][0] = 3;
dt[2].costs[1][1] = min_cost[2][1] = 1;
    dt[2].costs[2][2] = min_cost[2][2] = 0;
    dt[2].costs[3][3] = min_cost[2][3] = 2;
    rtpkt pki;
pki.sourceid = 2;
    for(i = 0; i < NODES; i++)
pki.mincost[i] = dt[2].costs[i][i];
    pki.destid = 0;
    tolayer2(pki);
    pki.destid = 1;
    tolayer2(pki);
    pki.destid = 3;
    tolayer2(pki);
spc2 += 3;
```

```c
}
/*-----------------------
 *    rtinit3
 *------------------- */
void rtinit3()
{
    int i;
for(i= 0; i < NODES; i+ + )
        for(int j= 0; j < NODES; j+ + )
dt[3].costs[i][j]= INF;
for(i= 0; i < NODES; i+ + )
min_cost[3][i]= INF;
    dt[3].costs[0][0]= min_cost[3][0]= 7;
    dt[3].costs[2][2]= min_cost[3][2]= 2;
    rtpkt pki;
pki.sourceid= 3;
    for(i= 0; i < NODES; i+ + )
pki.mincost[i]= dt[3].costs[i][i];
    pki.destid= 0;
    tolayer2(pki);
    pki.destid= 2;
    tolayer2(pki);
    spc3 + = 2;
}
/*-----------------------
 *    rtupdate0
 *------------------- */
void rtupdate0 (struct rtpkt *rcvdpkt)
{
    int i, newWeight;
    bool changed= false;       //check if the mincost has changed.
rtpkt pku;
    pku.sourceid= 0;
    for(i= 0; i < NODES; i+ + )
pku.mincost[i]= min_cost[0][i];    //initializing
    rpc0+ + ;
    for(i= 0; i < NODES; i+ + )//kernel part of the DVAlgorithm
    {
        newWeight= dt[0].costs[rcvdpkt- > sourceid][rcvdpkt- > sourceid] + rcvd-
```

```
                pkt->mincost[i];//receive a packet, update according to the incoming info.
                if(newWeight < dt[0].costs[i][rcvdpkt->sourceid] && i!=0)
                {
                    dt[0].costs[i][rcvdpkt->sourceid]= newWeight;//check if the private
                    costs table should be updated.
                    if(newWeight < min_cost[0][i])//check if the public min_cost table should
                    be updated
                    {
                        pku.mincost[i]= min_cost[0][i]= newWeight;
                        changed= true;
                    }
                }
            }
            if(changed){
                pku.destid= 1;
                tolayer2(pku);
                pku.destid= 2;
                tolayer2(pku);
                pku.destid= 3;
                tolayer2(pku);
                spc0+= 3;
            }
        }
        /*----------------------
         *    rtupdate1
         *---------------------- */
        void rtupdate1(struct rtpkt *rcvdpkt)
        {
            int i, newWeight;
            bool changed= false;
        rtpkt pku;
            pku.sourceid= 1;
            for(i= 0; i< NODES; i++)
                pku.mincost[i]= min_cost[1][i];
            rpc1++;
            for(i= 0; i< NODES; i++)
            {
                newWeight= dt[1].costs[rcvdpkt->sourceid][rcvdpkt->sourceid]+ rcvd-
                pkt->mincost[i];
```

```
            if(newWeight< dt[1].costs[i][rcvdpkt->sourceid] && i!=1)
            {
                    dt[1].costs[i][rcvdpkt->sourceid]= newWeight;
                    if(newWeight < min_cost[1][i])
                    {
                            pku.mincost[i]= min_cost[1][i]= newWeight;
                            changed= true;
                    }
            }
        }
        if(changed){
            pku.destid= 0;
            tolayer2(pku);
            pku.destid= 2;
            tolayer2(pku);
            spc1 += 2;
        }
}
/*----------------------
 *    rtupdate2
 *---------------------- */
void rtupdate2(struct rtpkt * rcvdpkt)
{
        int i, newWeight;
        bool changed= false;
rtpkt pku;
        pku.sourceid= 2;
        for(i= 0; i< NODES; i++)
            pku.mincost[i]= min_cost[2][i];

        rpc2++;
        for(i= 0; i< NODES; i++)
        {
            newWeight= dt[2].costs[rcvdpkt->sourceid][rcvdpkt->sourceid] + rcvd-
            pkt->mincost[i];
            if(newWeight< dt[2].costs[i][rcvdpkt->sourceid] && i!=2)
            {
                    dt[2].costs[i][rcvdpkt->sourceid]= newWeight;
                    if(newWeight < min_cost[2][i])
```

```
                {
                    pku.mincost[i]= min_cost[2][i]= newWeight;
                    changed= true;
                }
            }
        }
        if(changed){
            pku.destid= 0;
            tolayer2(pku);
            pku.destid= 1;
            tolayer2(pku);
            pku.destid= 3;
            tolayer2(pku);
            spc2+ = 3;
        }
}
/*-----------------------
 *    rtupdate3
 *----------------------- */
void rtupdate3(struct rtpkt * rcvdpkt)
{
        int i, newWeight;
        bool changed= false;
rtpkt pku;
        pku.sourceid= 3;
        for(i= 0; i< NODES; i+ + )
            pku.mincost[i]= min_cost[3][i];
        rpc3+ + ;
        for(i= 0; i< NODES; i+ + )
        {
            newWeight= dt[3].costs[rcvdpkt- > sourceid][rcvdpkt- > sourceid] + rcvd-
                pkt- > mincost[i];
if(newWeight< dt[3].costs[i][rcvdpkt- > sourceid] && i!= 3)
            {
                dt[3].costs[i][rcvdpkt- > sourceid]= newWeight;
                if(newWeight < min_cost[3][i])
                {
                    pku.mincost[i]= min_cost[3][i]= newWeight;
                    changed= true;
```

```
                }
            }
        }
        if(changed){
            pku.destid=0;
            tolayer2(pku);
            pku.destid=2;
            tolayer2(pku);
            spc3+=2;
        }
}
/*--------------------------------------------
 *      printdt0
 *
 *      Prints distance table at node 0.
 *-------------------------------------------- */
void printdt0 (struct distance_table *dtptr)
{
    printf("             via     \n");
    printf("   D0 |    1    2    3\n");
    printf("  ----|-----------------\n");
    printf("     1|  %3d   %3d  %3d\n",dtptr->costs[1][1],dtptr->costs[1][2],dtptr->costs[1][3]);
    printf("dest 2|  %3d   %3d  %3d\n",dtptr->costs[2][1],dtptr->costs[2][2],dtptr->costs[2][3]);
    printf("     3|  %3d   %3d  %3d\n",dtptr->costs[3][1],dtptr->costs[3][2],dtptr->costs[3][3]);
}
/*--------------------------------------------
 *      printdt1
 *
 *      Prints distance table at node 1.
 *-------------------------------------------- */
void printdt1 (struct distance_table *dtptr)
{
    printf("          via  \n");
    printf("   D1 |   0    2\n");
    printf("  ----|----------\n");
    printf("     0|  %3d   %3d\n",dtptr->costs[0][0], dtptr->costs[0][2]);
```

```
      printf("dest 2|    % 3d      % 3d\n",dtptr-> costs[2][0], dtptr-> costs[2][2]);
      printf("      3|   % 3d      % 3d\n",dtptr-> costs[3][0], dtptr-> costs[3][2]);
}

/*--------------------------------------------------
 *      printdt2
 *
 *      Prints distance table at node 2.
 *-------------------------------------------------*/
void printdt2 (struct distance_table * dtptr)
{
    printf("              via    \n");
    printf("   D2 |    0    1    3 \n");
    printf("   ----| ----------------\n");
    printf("      0|   % 3d   % 3d   % 3d\n",dtptr-> costs[0][0],dtptr-> costs[0][1], dtptr-> costs[0][3]);
    printf("dest 1|    % 3d   % 3d   % 3d\n",dtptr-> costs[1][0],dtptr-> costs[1][1],dtptr-> costs[1][3]);
    printf("      3|   % 3d   % 3d   % 3d\n",dtptr-> costs[3][0],dtptr-> costs[3][1], dtptr-> costs[3][3]);
}
/*--------------------------------------------------
 *      printdt3
 *
 *      Prints distance table at node 3.
 *-------------------------------------------------*/
void printdt3 (struct distance_table * dtptr)
{
    printf("              via    \n");
    printf("   D3 |    0    2 \n");
    printf("   ----| -----------\n");
    printf("      0|   % 3d    % 3d\n",dtptr-> costs[0][0], dtptr-> costs[0][2]);
    printf("dest 1|    % 3d   % 3d\n",dtptr-> costs[1][0], dtptr-> costs[1][2]);
    printf("      2|   % 3d    % 3d\n",dtptr-> costs[2][0], dtptr-> costs[2][2]);
}
/*--------------------------------------------------
 *      linkhandler0 & linkhandler1
 *-------------------------------------------------*/
void linkhandler0 (int linkid, int newcost) {}
```

```
void linkhandler1 (int linkid, int newcost) {}
```

3. DVsim.cpp 文件

```
//DVSim.cpp:
/***************************************************************  *
*       DVsim.c      Distance Vector Routing Simulation Package
*
*       Event driven simulation of distributed, asynchronous,
*       distance vector routing.  This file contains the network
*       simulation code which emulates layer 2 and below:  *
*            - simulates packet tranmissions between two
*              physically connected nodes
*
*            - assumes with no packet loss or corruption
*
*            - calls initializations routines rtinit0, etc., before
*              beginning simulation
*
*
*       Original code from Kurose and Ross, Computer Networking
*       1. Changed RNG to use drand48() with seed= getpid()
*       2. Replaced 999 with INF
*       3. created common include file DVsim.h
* ------------------------------------------------------------ */
#define         DVSIM_H
#include       "DVsim.h"
#define         FROM_LAYER2     2
#define         LINK_CHANGE     10
extern int rpc0;
extern int rpc1;
extern int rpc2;
extern int rpc3;
extern int spc0;
extern int spc1;
extern int spc2;
extern int spc3;
/* -----------------------------------
*       Event list and possible events
```

```c
 * --------------------------------------- */
struct event
{
    double      evtime;         /* event time */
    int         evtype;         /* event type code */
    int         eventity;       /* entity where event occurs */
    struct rtpkt *rtpktptr;     /* ptr to packet (if any) assoc w/ this event */
    struct event *prev;
    struct event *next;
};
struct event *evlist= NULL;
/* -------------------------------
 *      Simulation clock
 * ----------------------------- */
double      clocktime= 0.000;
double      get_clock() { return clocktime; }
/* -------------------------------
 *      Function prototypes
 * ----------------------------- */
void        init();
void        insertevent (struct event *);
void        printevlist();
void        tolayer2 (struct rtpkt );
/* -----------------------------
 *      MAIN
 * ----------------------------- */
int main()
{
    struct event *eventptr;
    int i;

    /* ---------------------------------------
     *      Loop until we run out of events
     * --------------------------------------- */
    init();
    printdt0(&dt[0]);
    printdt1(&dt[1]);
printdt2(&dt[2]);
    printdt3(&dt[3]);
```

```c
while (1)
{
    /*---------------------------------
     *   Get next event from top of list
     *      Done if event list is empty
     *--------------------------------- */
        if (!(eventptr= evlist)) break;
        evlist= evlist-> next;
        if (evlist) evlist-> prev= NULL;
    /*---------------------------------
     *   Trace info
     *--------------------------------- */
        if (TRACE> 1) {
            printf("\nMAIN: rcv event, t= %.3f, at %d",
                eventptr-> evtime, eventptr-> eventity);
            if (eventptr-> evtype = = FROM_LAYER2 ) {
            printf(" src:%2d,", eventptr-> rtpktptr-> sourceid);
printf(" dest:%2d,", eventptr-> rtpktptr-> destid);
printf(" contents: ");
            for (i= 0; i< NODES; i+ + ) printf ("%3d ",eventptr-> rtpktptr-> mincost
                [i]);
                printf("\n");
        }
        }
    /*---------------------------------
     *   Update clock to next event time
     *--------------------------------- */
        clocktime= eventptr-> evtime;


    /*---------------------------------
     *   Call event handler
     *--------------------------------- */
        if (eventptr-> evtype = = FROM_LAYER2 ) {
            if     (eventptr-> eventity = = 0) {rtupdate0(eventptr-> rtpk-
                tptr);printdt0(&dt[0]);}
            else if (eventptr-> eventity = = 1) {rtupdate1(eventptr-> rtpktptr);
                printdt1(&dt[1]);}
            else if (eventptr-> eventity = = 2) {rtupdate2(eventptr-> rtpktptr);
                printdt2(&dt[2]);}
```

```
                else if (eventptr->eventity = = 3) {rtupdate3(eventptr->rtpktptr);
                printdt3(&dt[3]);}

                else { printf("Panic: unknown event entity\n"); exit(0); }
        }
            else if (eventptr->evtype = = LINK_CHANGE ) {
if (clocktime <  10001.0) { linkhandler0(1,20); linkhandler1(0,20); }
            else                  { linkhandler0(1,1); linkhandler1(0,1); }
}
            else {
            printf("Panic: unknown event type\n");
            exit(0);
            }
    /*----------------------------------------------------
     *   Free memory for packet (if any) and event struct
     *------------------------------------------------- */
        if (eventptr->evtype = = FROM_LAYER2 ) free(eventptr->rtpktptr);
        free(eventptr);
    }
    /*------------------------------
     *     SIMULATION TERMINATED
     *---------------------------- */
    printf("\nSimulator terminated at t=%f, no packets in medium\n",clocktime);
    cout << "Node0 total sended " << spc0 << " packets and received " << rpc0 << " packets. " << endl;
    cout << "Node1 total sended " << spc1 << " packets and received " << rpc1 << " packets. " << endl;
    cout << "Node2 total sended " << spc2 << " packets and received " << rpc2 << " packets. " << endl;
    cout << "Node3 total sended " << spc3 << " packets and received " << rpc3 << " packets. " << endl;
//    return 0;
}
/*------------------------------
 *     init
 *---------------------------- */
void init()
{
    int i;
```

第13章 一个分布式异步距离向量算法

```c
    double    sum, avg;
    struct event *evptr;
  /*---------------------
   *    Input TRACE
   *------------------*/
    printf ("Enter TRACE:");
    scanf ("%d", &TRACE);
  /*---------------------
   *    Test RNG
   *------------------*/
    srand ((unsigned) getpid());
for (i=0, sum=0.; i< 1000; i++) sum+ = rand();
    avg= (sum / 1000. 0 - 15000) / 2000;
if (avg < 0.25 || avg > 0.75) {
        printf("RNG PROBLEM\n");
      printf("Average of 1000 random numbers is %g   (expected 0.5).\n", avg);
        printf("It is likely that random number generation on your machine\n");
        printf("is different from what this emulator expects. Please take\n");
        printf("a look at the routine init() in the emulator code. \n");
        exit(0);
    }
  /*------------------------------------
   *    Initialize clock and nodes
   *---------------------------------*/
    clocktime= 0.0;
    rtinit0();
    rtinit1();
    rtinit2();
    rtinit3();
  /*------------------------------------
   *    Initialize future link changes
   *---------------------------------*/
    if (LINKCHANGES = = 1)
    {
        evptr= (struct event *)malloc(sizeof(struct event));
        evptr-> evtime= 10000.0;
        evptr-> evtype= LINK_CHANGE;
        evptr-> eventity= -1;
        evptr-> rtpktptr= NULL;
```

```
            insertevent(evptr);
            evptr= (struct event * )malloc(sizeof(struct event));
            evptr-> evtype= LINK_CHANGE;
            evptr-> evtime= 20000.0;
            evptr-> rtpktptr= NULL;
            insertevent(evptr);
      }
}
/****************** EVENT HANDLING ROUTINES ******
*
*       These routines handle the event list
*
***************************************************/
/* -----------------------------------------------------------
*       insertevent
*
*             Inserts object in event list. List is sorted by time.
*       p -> object to be inserted
*       q -> top of list
* ----------------------------------------------------------- */
void insertevent (struct event * p)
{
    struct event * q,* qold;

    if (TRACE> 3) {
        printf("            INSERTEVENT: time is % lf\n",clocktime);
        printf("            INSERTEVENT: future time will be % lf\n",p-> evtime);
    }
    q= evlist;
    if (q= = NULL) {
      /* -----------------------------
             *    Empty list - insert at top
       * ----------------------------- */
        evlist= p;
        p-> next= NULL;
        p-> prev= NULL;
    }
    else {
      /* -------------------------------------
```

```
                    *   Find event time > p's time
      * ------------------------------------ */
         for(qold= q; q != NULL && p-> evtime > q-> evtime; q= q-> next) qold= q;

         if (q= = NULL) {
         /* --------------------
                    *   End of list
          * -------------------- */
            qold-> next= p;
            p-> prev= qold;
            p-> next= NULL;
         }
         else if (q= = evlist) {
         /* --------------------
                    *   Top of list
          * -------------------- */
            p-> next= evlist;
            p-> prev= NULL;
            p-> next-> prev= p;
            evlist= p;
         }
         else {
         /* --------------------
                    *   Middle of list
          * -------------------- */
            p-> next= q;
            p-> prev= q-> prev;
            q-> prev-> next= p;
            q-> prev= p;
         }
      }
}
/* ----------------------------------------------------
 *    printevlist - prints event list
 * ---------------------------------------------------- */
void printevlist()
{
   struct event * q;
   printf("--------------\nEvent List Follows:\n");
```

```
    for (q= evlist; q! = NULL; q= q-> next)
        printf("Event time: % f, type: % d entity: % d\n",q-> evtime,q-> evtype,q-> eventity);
        printf("--------------\n");
}
/*----------------------------------------------------------
 *     tolayer2
 *---------------------------------------------------------*/
void tolayer2 (struct rtpkt packet)
{
    struct rtpkt * mypktptr;
    struct event * evptr, * q;
    double     lastime;
    int        i;
/*------------------------------------------------------------
 * Check for packet errors
 * Check if source and destination id's are reasonable
 *-----------------------------------------------------------*/
    if (packet. sourceid< 0 || packet. sourceid > NODES) {
        printf("WARNING: illegal source id in your packet, ignoring packet! \n");
        return;
    }
    if (packet. destid< 0 || packet. destid > NODES) {
        printf("WARNING: illegal dest id in your packet, ignoring packet! \n");
        return;
    }
    if (packet. sourceid = = packet. destid) {
        printf("WARNING: source and destination id's the same, ignoring packet! \n");
        return;
    }
    if (linkcost[packet. sourceid][packet. destid] = = INF) {
        printf("WARNING: source and destination not connected, ignoring packet! \n");
        return;
    }
/*------------------------------------------------------------
 * Make a copy of the packet student just gave me since he/she may decide
 * to do something with the packet after we return back to him/her.
 *-----------------------------------------------------------*/
    mypktptr= (struct rtpkt * ) malloc(sizeof(struct rtpkt));
```

```c
mypktptr->sourceid= packet.sourceid;
mypktptr->destid  = packet.destid;
for (i=0; i<NODES; i++) mypktptr->mincost[i]= packet.mincost[i];
/*--------------------
 * Print trace info
 *------------------*/
if (TRACE>2)
{
    printf(" TOLAYER2: source: %d, dest: %d\n              costs:",
mypktptr->sourceid, mypktptr->destid);
    for (i=0; i<4; i++)   printf("%d ",mypktptr->mincost[i]);
printf("\n");
}
/*-----------------------------------------------------------------
 * Create future event for arrival of packet at the other side
 *------------------------------------------------------------- */
evptr           = (struct event *)malloc(sizeof(struct event));
evptr->evtype   = FROM_LAYER2;    /* packet will pop out from layer3 */
evptr->eventity= packet.destid;   /* event occurs at other entity */
evptr->rtpktptr= mypktptr;         /* save ptr to my copy of packet */
lastime= clocktime;
for (q=evlist; q!=NULL; q=q->next)
{
    if ( (q->evtype==FROM_LAYER2 && q->eventity==evptr->eventity) )
        lastime= q->evtime;
}
evptr->evtime= lastime + 2.*rand();
/*-----------------
 * Trace info
 *---------------- */
if (TRACE>2) printf("      TOLAYER2: scheduling arrival on other side\n");
/*---------------------------------------------
 * Insert message arrival into event list
 *----------------------------------------- */
insertevent(evptr);
}
```

第 14 章 RTSP 和 RTP 实现流媒体点播系统

14.1 实验目标

本实验使用 C/S 结构，完成一个流媒体点播的服务器端以及客户端点播的开发，使用 RTSP（real-time streaming protocol）完成流媒体点播的控制，使用 RTP（real-time transfer protocol）完成流媒体格式文件的数据传输。客户端可以对流媒体文件进行请求、播放、暂停和停止，对接收到的流媒体数据包进行解压，服务器端完成响应客户端的请求、对流媒体数据文件进行封装成包、发送数据包的功能。

14.2 系统设计与组成

系统由四个类构成，分别实现了不同的功能。
Client：使用 GUI 设计客户端界面、点播视频的基本操作按钮和视频播放界面；用 RTSP 封装客户端控制命令，如请求、暂停、播放、停止等。
Server：用于响应相应客户端的 RTSP 请求，发送流媒体文件数据包。
RTPpacket：用于处理 RTP 数据包，将流媒体文件进行封装和解封装。
VideoStream：用于从流媒体文件中读取数据，在客户端播放界面上播放。
如图 14-1 所示，首先开启服务器，并创建一个欢迎套接字监听指定端口，然后客户端通过"客户端 RTSPsocket"对服务器发起连接，利用 RTSPsocket 创建客户端 RTSPBufferReader、RTSPBufferWriter 对象和服务器端对应的 RTSPBufferReader、RTSPBufferWriter 对象，并建立它们之间对应的连接，客户端和服务器端通过 RTSPBufferReader 读取（接收）数据和 RTSPBufferWriter 写（发送）数据，进行请求和响应交互。使用 RTSP 进行控制命令的请求和响应，使用 RTP 传输流媒体文件数据。

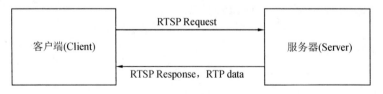

图 14-1 客户端服务器端设计

14.3 重要类及方法

1. Client 类中的重要类和方法

1）组合类
（1）timer，为接收到的数据包计时；
（2）rcvdp，接收从服务器端发送来的 UDP 数据包；
（3）RTPsocket，接收和发送 RTP 数据包；
（4）RTSPsocket，接收和发送 RTSP 数据包；
（5）RTSPBufferReader、RTSPBufferWriter，从 RTSPsocket 读取和向 RTSPsocket 写入数据。

2）内部类
setupButtonListener、playButtonListener、pauseButtonListener、tearButtonListener，分别监听客户端 GUI 上的连接、播放、暂停和停止四个功能按钮。

3）方法
parse_server_response（ ），解封装从服务器发来的数据，去掉首部信息，获取有效信息；send_RTSP_request（String request_type），封装客户端到服务器端发送的 RTSP 控制信息。

2. Server 类中的重要类和方法

1）组合类
timer，控制发送到客户端的数据的频率；
video，对流媒体文件进行操作；
RTPsocket，接收和发送 RTP 数据包；
senddp，将流媒体帧封装到 UDP 数据包中；
RTSPsocket，接收和发送 RTSP 数据包；
RTSPBufferReader、RTSPBufferWriter，从 RTSPsocket 读取和向 RTSPsocket 写入数据。

2）方法
actionPerformed（ActionEvent e），用 timer 控制取数据和发送数据的频率；parse_RTSP_request（ ），解封装客户端到服务器的 RTSP 控制信息；send_RTSP_response（ ），发送 RTSP 响应信息。

3. RTPpacket 类中的重要方法

方法：RTPpacket（int PType, int Framenb, int Time, byte [] data, int data_length），构造器，生成首部信息，将首部和播放文件的数据封装成 RTP-packet 包；RTPpacket（byte [] packet, int packet_size），构造器，解析一个 RTPpacket 包，得到它的有效信息和长度。

4. VideoStream 中的重要类与方法

1) 组合类

fis，文件的输入流类，用于从文件向运行程序的内存中取出数据。

2) 方法

getnextframe（byte [] frame），从流媒体文件中获取下一帧，并用数组返回正中的数据和帧的长度。

14.4 开 发 环 境

此系统使用 JAVA 语言开发，运行在 Windows 平台下，使用前应安装并配置好相关开发环境。

14.5 运 行 结 果

1. 运行步骤

（1）先运行服务器端程序，再运行客户端程序；

（2）在客户端的 Server、Port、File 栏中填入服务器地址、服务器端运行程序的端口号和要请求的文件（本实验中默认客户端与服务器端在一台主机，端口号为非知名端口 10001，文件为原书作者提供的流媒体文件 Movie.Mjpeg）；

（3）点击 Setup 进行与服务器连接，接着进行播放操作。

2. 运行结果

服务器端运行界面如图 14-2 所示。

图 14-2　服务端运行界面

客户端运行界结果如下。

未选择前如图 14-3 所示。

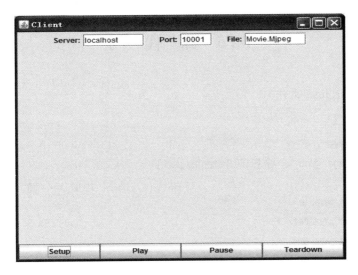

图 14-3　客户端未选择前界面

单击 Setup 后如图 14-4 所示。

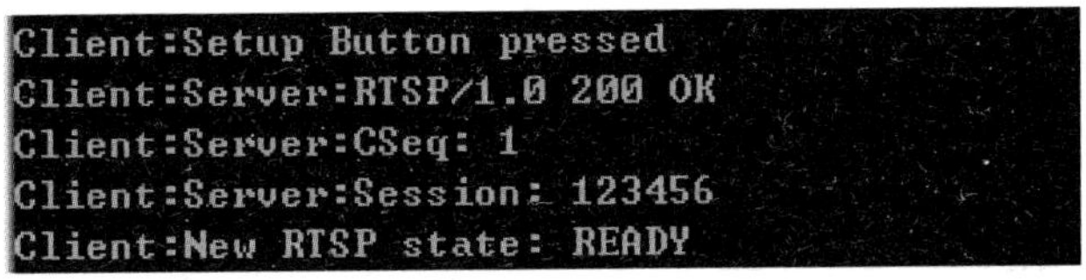

图 14-4　单击 Setup 后界面显示

单击 Play 后如图 14-5 所示。

图 14-5　单击 Play 后界面显示

14.6 源 代 码

1. Server.java 文件

```java
/*---------------------
    Server
    usage: java Server [RTSP listening port]
---------------------*/
import java.io.*;
import java.net.*;
import java.awt.*;
import java.util.*;
import java.awt.event.*;
import javax.swing.*;
import javax.swing.Timer;
public class Server extends JFrame implements ActionListener {
    // RTP variables:
    //----------------
    DatagramSocket RTPsocket; // socket to be used to send and receive UDP
                              // packets
    DatagramPacket senddp; // UDP packet containing the video frames
    InetAddress ClientIPAddr; // Client IP address
    int RTP_dest_port= 0; // destination port for RTP packets (given by the
                          // RTSP Client)
    // GUI:
    //----------------
    JLabel label;
    // Video variables:
    //----------------
    int imagenb= 0; // image nb of the image currently transmitted
    VideoStream video; // VideoStream object used to access video frames
    static int MJPEG_TYPE= 26; // RTP payload type for MJPEG video
    static int FRAME_PERIOD= 100; // Frame period of the video to stream, in ms
    static int VIDEO_LENGTH= 500; // length of the video in frames
    Timer timer; // timer used to send the images at the video frame rate
    byte[] buf; // buffer used to store the images to send to the client
    // RTSP variables
```

```java
// ----------------
// rtsp states
final static int INIT= 0;
final static int READY= 1;
final static int PLAYING= 2;
// rtsp message types
final static int SETUP= 3;
final static int PLAY= 4;
final static int PAUSE= 5;
final static int TEARDOWN= 6;
static int state; // RTSP Server state = = INIT or READY or PLAY
Socket RTSPsocket; // socket used to send/receive RTSP messages
// input and output stream filters
static BufferedReader RTSPBufferedReader;
static BufferedWriter RTSPBufferedWriter;
static String VideoFileName; // video file requested from the client
static int RTSP_ID= 123456; // ID of the RTSP session
int RTSPSeqNb= 0; // Sequence number of RTSP messages within the session
final static String CRLF= "\r\n";
// -------------------------------
// Constructor
// -------------------------------
public Server() {
    // init Frame
    super("Server");
    // init Timer
    timer= new Timer(FRAME_PERIOD, this);
    timer.setInitialDelay(0);
    timer.setCoalesce(true);
    // allocate memory for the sending buffer
    buf= new byte[15000];
    // Handler to close the main window
    addWindowListener(new WindowAdapter() {
        public void windowClosing(WindowEvent e) {
            // stop the timer and exit
            timer.stop();
            System.exit(0);
        }
    });
```

```java
        // GUI:
        label= new JLabel("Send frame #          ", JLabel.CENTER);
        getContentPane().add(label, BorderLayout.CENTER);
    }
    //----------------------------------
    // main
    //----------------------------------
    public static void main(String argv[]) throws Exception {
        // create a Server object
        Server theServer= new Server();
        // show GUI:
        theServer.pack();
        theServer.setVisible(true);
        // get RTSP socket port from the command line
        int RTSPport= 10001;
        //int RTSPport= Integer.parseInt(argv[0]);
        // Initiate TCP connection with the client for the RTSP session
        ServerSocket listenSocket= new ServerSocket(RTSPport);
        theServer.RTSPsocket= listenSocket.accept();
        listenSocket.close();
        // Get Client IP address
        theServer.ClientIPAddr= theServer.RTSPsocket.getInetAddress();
        // Initiate RTSPstate
        state= INIT;
        // Set input and output stream filters:
        RTSPBufferedReader= new BufferedReader(new InputStreamReader(
                theServer.RTSPsocket.getInputStream()));
        RTSPBufferedWriter= new BufferedWriter(new OutputStreamWriter(
                theServer.RTSPsocket.getOutputStream()));
        // Wait for the SETUP message from the client
        int request_type;
        boolean done= false;
        while (!done) {
            request_type= theServer.parse_RTSP_request(); // blocking
            if (request_type = = SETUP) {
                done= true;
                // update RTSP state
                state= READY;
                System.out.println("Server: New RTSP state: READY");
```

```java
            // Send response
            theServer.send_RTSP_response();
            // init the VideoStream object:
            theServer.video= new VideoStream(VideoFileName);
            // init RTP socket
            theServer.RTPsocket= new DatagramSocket();
        }
    }
    // loop to handle RTSP requests
    while (true) {
        // parse the request
        request_type= theServer.parse_RTSP_request(); // blocking
        if ((request_type = = PLAY) && (state = = READY)) {
            // send back response
            theServer.send_RTSP_response();
            // start timer
            theServer.timer.start();
            // update state
            state= PLAYING;
            System.out.println("Server: New RTSP state: PLAYING");
        } else if ((request_type = = PAUSE) && (state = = PLAYING)) {
            // send back response
            theServer.send_RTSP_response();
            // stop timer
            theServer.timer.stop();
            // update state
            state= READY;
            System.out.println("Server: New RTSP state: READY");
        } else if (request_type = = TEARDOWN) {
            // send back response
            theServer.send_RTSP_response();
            // stop timer
            theServer.timer.stop();
            // close sockets
            theServer.RTSPsocket.close();
            theServer.RTPsocket.close();
            System.exit(0);
        }
    }
```

```java
}
// ------------------------
// Handler for timer
// ------------------------
public void actionPerformed(ActionEvent e) {
    // if the current image nb is less than the length of the video
    if (imagenb < VIDEO_LENGTH) {
        // update current imagenb
        imagenb++;
        try {
            // get next frame to send from the video, as well as its size
            int image_length= video.getnextframe(buf);
            // Builds an RTPpacket object containing the frame
            RTPpacket rtp_packet= new RTPpacket(MJPEG_TYPE, imagenb,
                    imagenb * FRAME_PERIOD, buf, image_length);
            // get to total length of the full rtp packet to send
            int packet_length= rtp_packet.getlength();
            // retrieve the packet bitstream and store it in an array of
            // bytes
            byte[] packet_bits= new byte[packet_length];
            rtp_packet.getpacket(packet_bits);
            // send the packet as a DatagramPacket over the UDP socket
            senddp= new DatagramPacket(packet_bits, packet_length,
                    ClientIPAddr, RTP_dest_port);
            RTPsocket.send(senddp);
            // System.out.println("Server: Send frame #"+ imagenb);
            // print the header bitstream
            rtp_packet.printheader();
            // update GUI
            label.setText("Send frame #" + imagenb);
        } catch (Exception ex) {
            System.out.println("Server: Exception caught: " + ex);
            System.exit(0);
        }
    } else {
        // if we have reached the end of the video file, stop the timer
        timer.stop();
    }
}
```

```java
// --------------------------------
// Parse RTSP Request
// --------------------------------
private int parse_RTSP_request() {
    int request_type= -1;
    try {
        // parse request line and extract the request_type:
        String RequestLine= RTSPBufferedReader.readLine();
        // System.out.println("RTSP Server -Received from Client:");
        System.out.println("Server:" + RequestLine);
        StringTokenizer tokens= new StringTokenizer(RequestLine);
        String request_type_string= tokens.nextToken();
        // convert to request_type structure:
        if ((new String(request_type_string)).compareTo("SETUP") == 0)
            request_type= SETUP;
        else if ((new String(request_type_string)).compareTo("PLAY") == 0)
            request_type= PLAY;
        else if ((new String(request_type_string)).compareTo("PAUSE") == 0)
            request_type= PAUSE;
        else if ((new String(request_type_string)).compareTo("TEARDOWN") == 0)
            request_type= TEARDOWN;
        if (request_type == SETUP) {
            // extract VideoFileName from RequestLine
            VideoFileName= tokens.nextToken();
        }
        // parse the SeqNumLine and extract CSeq field
        String SeqNumLine= RTSPBufferedReader.readLine();
        System.out.println("Server:" + SeqNumLine);
        tokens= new StringTokenizer(SeqNumLine);
        tokens.nextToken();
        RTSPSeqNb= Integer.parseInt(tokens.nextToken());
        // get LastLine
        String LastLine= RTSPBufferedReader.readLine();
        System.out.println("Server:" + LastLine);
        if (request_type == SETUP) {
            // extract RTP_dest_port from LastLine
            tokens= new StringTokenizer(LastLine);
            for (int i= 0; i < 3; i++)
                tokens.nextToken(); // skip unused stuff
```

```
                    RTP_dest_port= Integer.parseInt(tokens.nextToken());
                }
                // else LastLine will be the SessionId line ... do not check for
                // now.
            } catch (Exception ex) {
                System.out.println("Server: Exception caught: " + ex);
                System.exit(0);
            }
        return (request_type);
    }
    //------------------------------------
    // Send RTSP Response
    //------------------------------------
    private void send_RTSP_response() {
        try {
            RTSPBufferedWriter.write("Server: RTSP/1.0 200 OK" + CRLF);
            RTSPBufferedWriter.write("Server: CSeq: " + RTSPSeqNb + CRLF);
            RTSPBufferedWriter.write("Server: Session: " + RTSP_ID + CRLF);
            RTSPBufferedWriter.flush();
            // System.out.println("RTSP Server - Sent response to Client.");
        } catch (Exception ex) {
            System.out.println("Server: Exception caught: " + ex);
            System.exit(0);
        }
    }
}
```

2. Client.java 文件

```
/*------------------
    Client
    usage: java Client [Server hostname] [Server RTSP listening port] [Video file requested]
    ---------------------- */

import java.io.*;
import java.net.*;
import java.util.*;
import java.awt.*;
```

```java
import java.awt.event.*;
import javax.swing.*;
import javax.swing.Timer;

public class Client{

    //GUI
    //----
    JFrame f= new JFrame("Client");
    JButton setupButton= new JButton("Setup");
    JButton playButton= new JButton("Play");
    JButton pauseButton= new JButton("Pause");
    JButton tearButton= new JButton("Teardown");
    JPanel mainPanel= new JPanel();
    JPanel buttonPanel= new JPanel();
    JLabel iconLabel= new JLabel();
    ImageIcon icon;


    //RTP variables:
    //---------------
    DatagramPacket rcvdp; //UDP packet received from the server
    DatagramSocket RTPsocket; //socket to be used to send and receive UDP packets
    static int RTP_RCV_PORT= 25000; //port where the client will receive the RTP packets

    Timer timer; //timer used to receive data from the UDP socket
    byte[] buf; //buffer used to store data received from the server

    //RTSP variables
    //---------------
    //rtsp states
    final static int INIT= 0;
    final static int READY= 1;
    final static int PLAYING= 2;
    static int state; //RTSP state == INIT or READY or PLAYING
    Socket RTSPsocket; //socket used to send/receive RTSP messages
    //input and output stream filters
    static BufferedReader RTSPBufferedReader;
    static BufferedWriter RTSPBufferedWriter;
```

```java
static String VideoFileName; //video file to request to the server
int RTSPSeqNb= 0; //Sequence number of RTSP messages within the session
int RTSPid= 0; //ID of the RTSP session (given by the RTSP Server)

final static String CRLF= "\r\n";

//Video constants:
//------------------
static int MJPEG_TYPE= 26; //RTP payload type for MJPEG video


//--------------------------
//Constructor
//--------------------------
public Client() {

    //build GUI
    // --------------------------

    //Frame
    f.addWindowListener(new WindowAdapter() {
        public void windowClosing(WindowEvent e) {
            System.exit(0);
        }
    });

    //Buttons
    buttonPanel.setLayout(new GridLayout(1,0));
    buttonPanel.add(setupButton);
    buttonPanel.add(playButton);
    buttonPanel.add(pauseButton);
    buttonPanel.add(tearButton);
    setupButton.addActionListener(new setupButtonListener());
    playButton.addActionListener(new playButtonListener());
    pauseButton.addActionListener(new pauseButtonListener());
    tearButton.addActionListener(new tearButtonListener());

    //Image display label
    iconLabel.setIcon(null);
```

```java
        //frame layout
        mainPanel.setLayout(null);
        mainPanel.add(iconLabel);
        mainPanel.add(buttonPanel);
        iconLabel.setBounds(0,0,380,280);
        buttonPanel.setBounds(0,280,380,50);

        f.getContentPane().add(mainPanel, BorderLayout.CENTER);
        f.setSize(new Dimension(390,370));
        f.setVisible(true);

        //init timer
        //--------------------------
        timer= new Timer(20, new timerListener());
        timer.setInitialDelay(0);
        timer.setCoalesce(true);

        //allocate enough memory for the buffer used to receive data from the server
        buf= new byte[15000];
    }


    //----------------------------------
    //main
    //----------------------------------
    public static void main(String argv[]) throws Exception
    {
        //Create a Client object
        Client theClient= new Client();

        //get server RTSP port and IP address from the command line
        //-----------------
        int RTSP_server_port= Integer.parseInt(argv[1]);
        String ServerHost= argv[0];
        InetAddress ServerIPAddr= InetAddress.getByName(ServerHost);

        //get video filename to request:
        VideoFileName= argv[2];

        //Establish a TCP connection with the server to exchange RTSP messages
```

```java
//------------------
    theClient.RTSPsocket= new Socket(ServerIPAddr, RTSP_server_port);

    //Set input and output stream filters:
    RTSPBufferedReader= new  BufferedReader(new InputStreamReader(theClient.RTSP-socket.getInputStream()) );
    RTSPBufferedWriter= new  BufferedWriter(new OutputStreamWriter(theClient.RTSP-socket.getOutputStream()) );

    //init RTSP state:
    state= INIT;
}


//------------------------------------
//Handler for buttons
//------------------------------------


//..............
//TO COMPLETE
//..............

//Handler for Setup button
//-----------------------
class setupButtonListener implements ActionListener{
    public void actionPerformed(ActionEvent e){

        //System.out.println("Setup Button pressed !");

        if (state = = INIT)
        {
            //Init non-blocking RTPsocket that will be used to receive data
            try{
                //construct a new DatagramSocket to receive RTP packets from the server, on port RTP_RCV_PORT
                RTPsocket= new DatagramSocket(RTP_RCV_PORT);

                //set TimeOut value of the socket to 5msec.
                RTPsocket.setSoTimeout(5);
```

```
        }
    catch (SocketException se)
    {
        System.out.println("Socket exception: "+ se);
        System.exit(0);
    }

//init RTSP sequence number
RTSPSeqNb= 1;

//Send SETUP message to the server
send_RTSP_request("SETUP");

//Wait for the response
if (parse_server_response() ! = 200)
    System.out.println("Invalid Server Response");
else
    {
        //change RTSP state and print new state
        state= READY;
        //System.out.println("New RTSP state: ....");
    }
    }//else if state ! = INIT then do nothing
  }
}

//Handler for Play button
//----------------------
class playButtonListener implements ActionListener {
    public void actionPerformed(ActionEvent e){

        //System.out.println("Play Button pressed !");

        if (state = = READY)
        {
            //increase RTSP sequence number
            RTSPSeqNb= RTSPSeqNb+ + ;
```

```
            //Send PLAY message to the server
            send_RTSP_request("PLAY");

            //Wait for the response
            if (parse_server_response() ! = 200)
                  System.out.println("Invalid Server Response");
            else
             {
                //change RTSP state and print out new state
                state= PLAYING;
                // System.out.println("New RTSP state: ...")

                //start the timer
                timer.start();
             }
        }//else if state ! = READY then do nothing
    }
}


//Handler for Pause button
// ----------------------
class pauseButtonListener implements ActionListener {
   public void actionPerformed(ActionEvent e){

        //System.out.println("Pause Button pressed !");

        if (state = = PLAYING)
          {
             //increase RTSP sequence number
        RTSPSeqNb= RTSPSeqNb+ + ;

            //Send PAUSE message to the server
            send_RTSP_request("PAUSE");

            //Wait for the response
            if (parse_server_response() ! = 200)
               System.out.println("Invalid Server Response");
```

```java
        else
         {
          //change RTSP state and print out new state
          state= READY;
          //System.out.println("New RTSP state: ...");

          //stop the timer
          timer.stop();
        }
      }
     //else if state ! = PLAYING then do nothing
   }
}

//Handler for Teardown button
// ----------------------
class tearButtonListener implements ActionListener {
   public void actionPerformed(ActionEvent e){

      //System.out.println("Teardown Button pressed !");

      //increase RTSP sequence number
      RTSPSeqNb= RTSPSeqNb+ + ;


      //Send TEARDOWN message to the server
      send_RTSP_request("TEARDOWN");

      //Wait for the response
      if (parse_server_response() ! = 200)
         System.out.println("Invalid Server Response");
      else
         {
          //change RTSP state and print out new state
          state= INIT;
          //System.out.println("New RTSP state: ...");

          //stop the timer
          timer.stop();
```

```
            //exit
            System.exit(0);
          }
        }
    }


// -----------------------------------
//Handler for timer
// -----------------------------------

class timerListener implements ActionListener {
    public void actionPerformed(ActionEvent e)   {

        //Construct a DatagramPacket to receive data from the UDP socket
        rcvdp= new DatagramPacket(buf, buf.length);

        try{
        //receive the DP from the socket:
        RTPsocket.receive(rcvdp);

        //create an RTPpacket object from the DP
        RTPpacket rtp_packet= new RTPpacket(rcvdp.getData(), rcvdp.getLength());

        //print important header fields of the RTP packet received:
        System.out.println("Got           RTP           packet         with         SeqNum         # "+ rtp_packet.getsequencenumber()+ " TimeStamp "+ rtp_packet.gettimestamp()+ " ms, of type "+ rtp_packet.getpayloadtype());

        //print header bitstream:
        rtp_packet.printheader();

        //get the payload bitstream from the RTPpacket object
        int payload_length= rtp_packet.getpayload_length();
        byte[] payload= new byte[payload_length];
        rtp_packet.getpayload(payload);

        //get an Image object from the payload bitstream
```

```java
            Toolkit toolkit= Toolkit.getDefaultToolkit();
            Image image= toolkit.createImage(payload, 0, payload_length);

            //display the image as an ImageIcon object
            icon= new ImageIcon(image);
            iconLabel.setIcon(icon);
        }
        catch (InterruptedIOException iioe){
            //System.out.println("Nothing to read");
        }
        catch (IOException ioe) {
            System.out.println("Exception caught: "+ ioe);
        }
    }
}


// ----------------------------------
//Parse Server Response
// ----------------------------------
private int parse_server_response()
{
    int reply_code= 0;

try{
        //parse status line and extract the reply_code:
        String StatusLine= RTSPBufferedReader.readLine();
        //System.out.println("RTSP Client - Received from Server:");
        System.out.println(StatusLine);

        StringTokenizer tokens= new StringTokenizer(StatusLine);
        tokens.nextToken(); //skip over the RTSP version
        reply_code= Integer.parseInt(tokens.nextToken());

        //if reply code is OK get and print the 2 other lines
        if (reply_code = = 200)
          {
            String SeqNumLine= RTSPBufferedReader.readLine();
            System.out.println(SeqNumLine);
```

```
                String SessionLine= RTSPBufferedReader.readLine();
                System.out.println(SessionLine);

                //if state = = INIT gets the Session Id from the SessionLine
                tokens= new StringTokenizer(SessionLine);
                tokens.nextToken(); //skip over the Session:
                RTSPid= Integer.parseInt(tokens.nextToken());
            }
        }
    catch(Exception ex)
        {
            System.out.println("Exception caught c: "+ ex);
            System.exit(0);
        }

    return(reply_code);
    }


// ----------------------------------
//Send RTSP Request
// ----------------------------------


//---------------------------------
//TO COMPLETE
//---------------------------------


private void send_RTSP_request(String request_type)
{
try{
        //Use the RTSPBufferedWriter to write to the RTSP socket

        //write the request line:
        RTSPBufferedWriter.write(request_type+ " "+ VideoFileName+ " RTSP/1.0"+ CRLF);
        System.out.print(request_type+ " "+ VideoFileName+ " RTSP/1.0"+ CRLF);

        //write the CSeq line:
        RTSPBufferedWriter.write("CSeq "+ RTSPSeqNb + CRLF);
        System.out.print("CSeq "+ RTSPSeqNb + CRLF);
```

```java
        //check if request_type is equal to "SETUP" and in this case write the Transport: line
        advertising to the server the port used to receive the RTP packets RTP_RCV_PORT
            if(request_type.equals("SETUP"))
            {
                RTSPBufferedWriter.write("Transport: RTP/UDP; client_port= "+ RTP_RCV_PORT + CRLF);
                System.out.print("Transport: RTP/UDP; client_port= "+ RTP_RCV_PORT + CRLF);
            }
            //otherwise, write the Session line from the RTSPid field
            else
            {
                RTSPBufferedWriter.write("Session: "+ RTSPid + CRLF);
                System.out.print("Session: "+ RTSPid + CRLF);
            }
            try
            {
            RTSPBufferedWriter.flush();
            }
            catch(IOException e)
            {
                System.out.println("Exception caught a: "+ e.toString());
            }
    }
    catch(Exception ex)
        {
            System.out.println("Exception caught b: "+ ex);
            System.exit(0);
        }
    }
}//end of Class Client
```

3. RTPpacket.java 文件

```java
import java.awt.Frame;
//class RTPpacket
public class RTPpacket {
    // size of the RTP header:
    static int HEADER_SIZE= 12;
```

```java
// Fields that compose the RTP header
public int Version;
public int Padding;
public int Extension;
public int CC;
public int Marker;
public int PayloadType;
public int SequenceNumber;
public int TimeStamp;
public int Ssrc;
// Bitstream of the RTP header
public byte[] header;
// size of the RTP payload
public int payload_size;
// Bitstream of the RTP payload
public byte[] payload;
//--------------------------
// Constructor of an RTPpacket object from header fields and payload
// bitstream
//--------------------------
public RTPpacket(int PType, int Framenb, int Time, byte[] data,
        int data_length) {
    // fill by default header fields:
    Version= 2;
    Padding= 0;
    Extension= 0;
    CC= 0;
    Marker= 0;
    Ssrc= 0;
    // fill changing header fields:
    SequenceNumber= Framenb;
    TimeStamp= Time;
    PayloadType= PType;
    // build the header bistream:
    //--------------------------
    header= new byte[HEADER_SIZE];
    //--------------------------
    // TO COMPLETE
    //--------------------------
```

```
// fill the header array of byte with RTP header fields
// header[0]= ---
//------
header[0]= (byte) (header[0] | 1 << 7);
header[1]= (byte) 26;
header[2]= (byte) (Framenb >> 8);
header[3]= (byte) (Framenb & 0xFF);
header[4]= (byte) (Time >> 24);
header[5]= (byte) ((Time & 0xFFFFFF) >> 16);
header[6]= (byte) ((Time & 0xFFFF) >> 8);
header[7]= (byte) (Time & 0xFF);
header[8]= (byte) 1;
header[9]= (byte) 1;
header[10]= (byte) 1;
header[11]= (byte) 1;
// fill the payload bitstream:
//-------------------------
payload_size= data_length;
payload= new byte[data_length];
// fill payload array of byte from data (given in parameter of the
// constructor)
//------
for (int i= 0; i < payload_size; i++ ){
    payload[i]= data[i];
}


printheader();
// ! Do not forget to uncomment method printheader() below !
}
//------------------------
// Constructor of an RTPpacket object from the packet bistream
//------------------------
public RTPpacket(byte[] packet, int packet_size){
    // fill default fields:
    Version= 2;
    Padding= 0;
    Extension= 0;
    CC= 0;
    Marker= 0;
```

```java
        Ssrc= 0;
        // check if total packet size is lower than the header size
        if (packet_size >= HEADER_SIZE) {
            // get the header bitsream;
            header= new byte[HEADER_SIZE];
            for (int i= 0; i < HEADER_SIZE; i++)
                header[i]= packet[i];
            // get the payload bitstream;
            payload_size= packet_size - HEADER_SIZE;
            payload= new byte[payload_size];
            for (int i= HEADER_SIZE; i < packet_size; i++)
                payload[i - HEADER_SIZE]= packet[i];
            // interpret the changing fields of the header;
            PayloadType= header[1] & 127;
            SequenceNumber= unsigned_int(header[3]) + 256
                * unsigned_int(header[2]);
            TimeStamp= unsigned_int(header[7]) + 256 * unsigned_int(header[6])
                + 65536 * unsigned_int(header[5]) + 16777216
                * unsigned_int(header[4]);
        }
    }
    //------------------------
    // getpayload: return the payload bistream of the RTPpacket and its size
    //------------------------
    public int getpayload(byte[] data) {
        for (int i= 0; i < payload_size; i++)
            data[i]= payload[i];
        return (payload_size);
    }
    //------------------------
    // getpayload_length: return the length of the payload
    //------------------------
    public int getpayload_length() {
        return (payload_size);
    }
    //------------------------
    // getlength: return the total length of the RTP packet
    //------------------------
    public int getlength() {
```

```java
        return (payload_size + HEADER_SIZE);
    }
    //-------------------------
    // getpacket: returns the packet bitstream and its length
    //-------------------------
    public int getpacket(byte[] packet) {
        // construct the packet= header + payload
        for (int i= 0; i < HEADER_SIZE; i++ )
            packet[i]= header[i];
        for (int i= 0; i < payload_size; i++ )
            packet[i + HEADER_SIZE]= payload[i];
        // return total size of the packet
        return (payload_size + HEADER_SIZE);
    }
    //-------------------------
    // gettimestamp
    //-------------------------
    public int gettimestamp() {
        return (TimeStamp);
    }
    //-------------------------
    // getsequencenumber
    //-------------------------
    public int getsequencenumber() {
        return (SequenceNumber);
    }
    //-------------------------
    // getpayloadtype
    //-------------------------
    public int getpayloadtype() {
        return (PayloadType);
    }
    //-------------------------
    // print headers without the SSRC
    //-------------------------
    public void printheader() {
        // TO DO: uncomment
        for (int i= 0; i < (HEADER_SIZE - 4); i++ ){
            for (int j= 7; j > = 0; j--)
```

```
                if (((1 < < j) & header[i]) ! = 0)
                    System. out. print("1");
                else
                    System. out. print("0");
                System. out. print(" ");
            }
            System. out. println();
        }
        // return the unsigned value of 8- bit integer nb
        static int unsigned_int(int nb) {
            if (nb > = 0)
                return (nb);
            else
                return (256 + nb);
        }
    }
```

4. VideoStream. java 文件

```
//VideoStream
import java. io. *;
public class VideoStream {
    FileInputStream fis; // video file
    int frame_nb; // current frame nb
    // -----------------------------------
    // constructor
    // -----------------------------------
    public VideoStream(String filename) throws Exception {
        // init variables
        fis= new FileInputStream(filename);
        frame_nb= 0;
    }
    public int getnextframe(byte[] frame) throws Exception {
        int length= 0;
        String length_string;
        byte[] frame_length= new byte[5];
        // read current frame length
        fis. read(frame_length, 0, 5);
        for(int i= 0; i < frame_length. length; i+ + )
```

```
        System.out.println(frame_length[i]);
    // transform frame_length to integer
    length_string= new String(frame_length);
    System.out.println(length_string);
    length= Integer.parseInt(length_string);
    return (fis.read(frame, 0, length));
    }
}
```

参 考 文 献

曹晓军. 2012. 计算机网络. 北京:科学出版社.
冯博琴,夏秦. 2008. 计算机网络实验教程. 北京:高等教育出版社.
胡宏智. 2013. 大学计算机基础实验与指导. 北京:高等教育出版社.
胡亮,徐高潮,魏晓辉. 2008. 计算机网络. 2版. 北京:高等教育出版社.
钱德沛. 2005. 计算机网络实验教程. 北京:高等教育出版社.
尚晓航. 2008. 计算机网络技术基础. 3版. 北京:高等教育出版社.
汤子瀛,哲凤屏,汤小丹. 2004. 计算机网络技术及其应用. 2版. 北京:电子工业出版社.
谢希仁. 2008. 计算机网络. 5版. 北京:电子工业出版社.
张曾科. 2009. 计算机网络. 北京:人民邮电出版社.
Andrew S T, David J W. 2010. Computer Networking. 5th ed. Boston:Pearson Education.
James F K, Keith W R. 2012. Computer Networking:A Top-Down Approach. 6th ed. Boston:Pearson Education.